INTRODUCTORY CLASSICAL MECHANICS

ABOUT THE AUTHORS

Mr. Juber Hawaldar

He was born in the year of 1998 at Bhalki, Bidar district and completed his primary schooling education from Class 1 to Class 10 at Guruprasad High School Bhalki – 585328. He completed his Pre – University Course (PUC) at Diamond PU Science College Bhalki – 585328. He has obtained his Bachelor of Science (B.Sc.) and Master of Science in Physics (M.Sc. in Physics) from Sri Channabasaveshwar Arts, Commerce and Science Degree College & Post Graduate Studies & Research Center Bhalki – 585328, which comes under Gulbarga University, Kalaburagi. He has started his teaching career as an Assistant Teacher at Red Rose Bunch Higher Primary School Bhalki (2019 – 2022). He is also well known Tutor. He is currently working as a lecturer in Physics at Sri Channabasaveshwar Post Graduate Studies and Research Center Bhalki - 585328. His well – known Published Books are "Science Concept & Last 5 Years Solved Papers (Chemistry)", "Science Concept & Last 5 Years Solved Papers (Biology)", "Basic Maths Book", "Monkey Drawing Book" , Rhymes Book for Children" and "JH Books Science Class – 10 (ISBN – 9798892330695)".

Dr. Santoshkumar Siddanna Teerthe

He was born in the year of 1983 at Padasavali Village of Aland taluka, Kalaburagi district and completed his primary to PUC Science education in the years 1992-2002 in his native village and Aland taluka. Also he was completed his B.Sc., B.Ed. and M.Sc. in Physics under Gulbarga University, Kalaburagi in the years 2002-2012 and further he was awarded as a Doctor of Philosophy in Physics at 2019 (Ph.D. in Physics). Also awarded as a "Young Scientist Award', "UGC – BSR fellow Award", "Karnataka state Adharsh Sikshaka Award" and "Appa Award". He has published more than 30 research papers at National and International research journals and supervised more than 5 different M.Sc. level projects. His research, teaching and Social work services are under process at National and International level. He has been involved in the teaching profession since 2002 onwards till today and he is currently working as a permanent Assistant Professor in Physics at Sri Channabasaveshwar Arts, Commerce and Science Degree College and PG research centre, Bhalki, Bidar district. He has always contributed to the welfare of students and society continues to do so.

INTRODUCTORY CLASSICAL MECHANICS

PART-I

Author

Mr. Juber Hawaldar

M.Sc., B.Ed.

Lecturer

Department of Physics

Sri. Channabasaveshwar Post Graduate Studies

& Research Center Bhalki – 585328

Co-author

Dr. Santoshkumar S. Teerthe

M.Sc., Ph.D.

Vice – Principal

&

Head of the Department & Asst. Professor

Department of Physics

Sri. Channabasaveshwar Arts, Commerce &

Science Degree College Bhalki – 585328

ACADEMIC ENCLAVE

Edited By

Mr. PASHUSAB N. *(M.Sc., B.Ed.)*
Head, Department of Physics
Sri. Channabasaveshwar Post Graduate Studies
& Research Center Bhalki – 585328

Mr. VISHWJIT MANKARI *(M.Sc., B.Ed.)*
Lecturer, Department of Physics
Sri. Channabasaveshwar Post Graduate Studies
& Research Center Bhalki – 585328

Mr. SYED JAFAR SIDDIQE ALI *(M.Sc., B.Ed.)*
Lecturer, Department of Mathematics
Diamond P.U. Science College Bhalki – 585328

Proof Reading

Prof. MEENAKUMARI PATIL *(M.Com. M.Phil. M.A., M.Ed.)*
Principal & Assistant Professor
Department of Commerce
Sri. Channabasaveshwar Post Graduate Studies
& Research Center Bhalki – 585328

Mrs. SABA KOUSER *(B.A., B.Ed.)*
Head Mistress
Red Rose Bunch School Bhalki - 585328

First Edition, 2024

ISBN: 978-81-971898-9-0 (Paperback)

Published by
ACADEMIC ENCLAVE
House No. 10/15, Ground Floor,
Block-A, Street No. 14/3, Kamal Vihar,
Kamal Pur, Burari, Delhi-110084
Email: info@academicenclave.com
Mobile: +91-7042423166
Website: www.academicenclave.com

DEDICATION

This Book is dedicated to my Parents because of their efforts and Sacrifice I became an educated and written this book.

This Book is dedicated to my Teachers, who gave me knowledge about alphabets and enhanced my skills.

This Book is dedicated to my Family, Relatives and Friends who encourages me at every time. Also this Book dedicated to my all Supporters and Well-wishers.

Foreword

In the vast and intricate landscape of physics, classical mechanics stands as one of the most fundamental and enduring pillars. It serves as the gateway through which we begin our journey into the world of physical science, offering profound insights into the laws that govern the motion of objects and the forces that influence them. It is with great enthusiasm that I introduce Introductory Classical Mechanics, a book that I believe will serve as an invaluable resource for students and enthusiasts alike.

Classical mechanics, though centuries old, continues to be a field of vibrant study and application. From the elegantly simple laws formulated by Newton to the more complex dynamics explored in modern physics, the principles of classical mechanics form the bedrock upon which much of contemporary science is built. This book does not merely recount these principles but seeks to illuminate their relevance and application in a manner that is both accessible and engaging.

The author, Mr. Juber Hawaldar, has crafted this text with a deep understanding of both the subject matter and the pedagogical challenges involved in teaching it. With clarity and precision, Mr. Juber Hawaldar presents the core concepts of classical mechanics, breaking down complex ideas into manageable and comprehensible segments.

As you embark on this journey through Introductory Classical Mechanics, I encourage you to approach the material with curiosity and an open mind. The principles you will encounter are not merely academic exercises but are key to unlocking a greater understanding of the physical world around us. Whether you are a student seeking a solid foundation or an enthusiast eager to explore the elegance of classical mechanics, this book will serve as a valuable companion in your quest for knowledge.

With anticipation for the discoveries that await you,

Sincerely,

Dr. Md. Azeem

M.A, M.Ed., M.Phil., Ph.D.
Assistant Professor
Manuu College of Teacher
Education Bidar - 585401

Acknowledgments

The journey of bringing Introductory Classical Mechanics from concept to completion has been a collaborative effort, and I am deeply grateful to all those who have supported and guided me along the way.

First and foremost, I want to express my heartfelt gratitude to the dedicated team at Academic Enclave for their unwavering support and commitment to this Book. Their guidance, patience, and tireless efforts have been instrumental in turning the vision of this book into a reality.

I also want to express my heartfelt gratitude to the Principal and Staff members of "Sri Channabasaveshwar Post Graduate Studies and Research Centre Bhalki" for their valuable suggestions and support to this book.

I would like to extend my heartfelt thanks to my mentors and colleagues, whose insightful feedback and encouragement have been invaluable. Your expertise and constructive criticism have not only helped refine the content of this book but have also inspired me to approach classical mechanics with greater depth and clarity.

I also want to express my heartfelt gratitude to Dr. Santoshkumar Teerthe, Dr. Md. Azeem, Dr. Kauthale Ramling, Prof. Meenakumari Patil, Mr. PashuSab N, Mr. Vishwajeet M, Mrs. Jyothi D. Anandwade, Mr. Arunkumar Mahajan, Miss. Anjali Kale, Miss. Ashmita Nirgude, Miss. Tejashri, Mr. Sharanappa, Mr. Arvind, Mr. Syed Jafar Siddiqe Ali, Mr. Syed Asif Ali, Mr. Shaik Musa, Mrs. Saba Kouser, Mr. Sidram Tamasange and Mr. Udhav Bhalke for their valuable suggestions to this book.

I am particularly indebted to my students, whose curiosity and enthusiasm have continually motivated me to explore new ways of presenting complex ideas. Your questions and insights have played a crucial role in shaping the approach and examples found in these pages.

Special thanks go to Dr. Santoshkumar Teerthe, who provided indispensable editorial support and ensured that the book's content is both accurate and engaging. Your meticulous attention to detail has greatly enhanced the quality of this work.

Finally, I would like to thank you, the reader, for choosing to embark on this exploration of classical mechanics. It is my hope that this book serves as a valuable resource and inspires a deeper appreciation for the beauty and relevance of the principles that govern our physical world.

Mr. Juber Hawaldar

Preface

It gives us great pleasure in bringing out the First Edition of this book.

Welcome to Introductory Classical Mechanics. This book aims to provide a clear, comprehensive, and engaging introduction to the fundamental principles of classical mechanics, a cornerstone of physical science.

Classical mechanics is the study of the motion of bodies under the influence of forces. It forms the foundation for understanding a wide range of physical phenomena, from the orbits of planets to the motion of everyday objects. Despite being a well-established field with principles that have stood the test of time, classical mechanics continues to be relevant, offering profound insights into both familiar and novel problems.

This book is designed for students who are new to the subject and seek a solid grounding in the basics of mechanics. It is also intended for those who wish to revisit and reinforce their understanding of the principles that govern the physical world. Whether you are a student in physics, engineering, or any related field, this book aims to support your learning journey through clear explanations.

The text is organized into several chapters, each focusing on key concepts and techniques in classical mechanics. We begin with the basics of kinematics and dynamics, gradually advancing to more complex topics such as systems of particles, rigid body dynamics, and oscillatory motion. Each chapter builds upon the previous ones, ensuring a coherent and progressive learning experience.

We look forward for the suggestions from the readers for the improvement of the book.

- Authors

Contents

CHAPTER – 1

INTRODUCTION TO NEWTONIAN MECHANICS

Father of Classical Mechanics

Isaac Newton is popularly remembered as the man who saw an apple fall from a tree, and was inspired to invent the theory of gravity. If you have grappled with elementary physics then you know that he invented calculus and the three laws of motion upon which all of mechanics is based.

1.1. INTRODUCTION

Definition:

"A branch of Physical science that deals with study of relation between energy and forces and their effects on bodies."

Classical mechanics deals with the motion of Physical bodies at the macroscopic level. Galileo and Sir Isaac Newton laid its foundation in the 17th century. As Newton's laws of motion provide the basis of classical mechanics, it is often referred to as *Newtonian mechanics.* There are two parts in mechanics: *Kinematics* and *Dynamics*. Kinematics deals with the geometrical description of the motion of objects without considering the forces producing the motion. Dynamics is the part that concerns the forces that produce changes in motion or the changes in other physical properties. This leads us to the concept of force, mass and laws that govern the motion of objects. To apply the laws to different situations, Newtonian mechanics has since been reformulated in a few different forms, such as the langrage, the Hamilton – Jacobi formalisms. All these formalisms are equivalent and their applications to topics of interests form the basis of this book.

Classical mechanics is the study of the motion of bodies under the action of physical forces. A force is any influence that can cause an object to change its velocity. The object can be anything from an elementary particle to a galaxy. Of course anything larger than an elementary particle is ultimately a composite of elementary particles, but fortunately we usually don't have to consider all those, and can coarse-grain to the scale of the objects at hand. As is true for any Physical model, classical mechanics is an approximation and has its limits - it breaks down at very small scales, high speeds and large gravitational fields - but within its range of applicability (which includes pretty much every single phenomenon in everyday life) it is extremely useful.

Classical mechanics is based on a small number of Physical laws, which are mathematical formulations of a physical observation. Some laws can be derived from others, but you cannot derive all of them from scratch. Some laws are axioms, and we'll assume they are valid. The laws we'll encounter can be divided up in three classes: Newton's laws of motion, conservation laws and force laws. As we'll see, the three conservation laws of classical mechanics (of energy, momentum and angular momentum) can be derived from Newton's second and third laws of motion, as can Newton's first law. The force laws give us the force exerted by a certain physical system - a compressed spring (Hooke's law) or two charged particles (Coulomb's law) for example. These also feed back into Newton's laws of motion, although they cannot be derived from these and are axioms by themselves.

In addition to the physical laws, there is a large number of definitions - which should not be confused with the laws. Definitions are merely convenient choices. A good example is the definition of the number π: half the ratio of the circumference to the radius of a circle. As you have no doubt noticed, it is very convenient that this number has gotten its own symbol that is universally recognized, as it pops up pretty much everywhere. However, there is no axiom here, as we are simply taking a ratio and giving it a name.

Classical mechanics is used for describing the motion of macroscopic objects, from projectiles to parts of machinery, as well as astronomical objects, such as spacecraft, planets, stars, and galaxies, etc.

Force:

Force can be defined as a push or a pull that changes or tends to change the state of rest or uniform motion of an object or changes the direction or shape of an object. It causes objects to accelerate or add to their overall pressure.

- Force is a vector quantity.
- S. I unit is Newton (N)
- Formula F = ma

  Where F = Force

  m = Mass of an object

  a = Acceleration

Momentum:

Momentum is Product of the mass of a object and its velocity.

- Momentum is a vector quantity; i.e., it has both magnitude and direction.
- S.I unit Kg ms^{-1} (Kilogram meter per second)
- Formula P = mv

  Where P = Momentum

  m = Mass of an object

  v = Velocity of an object

Aristotle Fallacy (322 BC – 384 BC)

Aristotle made observations from practical experiences and came to the conclusion that an external force is required to keep a body in uniform motion. This is known as Aristotle's fallacy.

Galileo Experiment (1564 – 1642)

Galileo observed that a heavy object and a light object will fall to the ground simultaneously from a particular height. All bodies of any mass will fall down simultaneously at the same time from a particular height. That is the acceleration of all free-falling bodies are same.

There are some fundamental experiments of Galileo in physical science given below:

Experiment of a free-falling body (Pisa tower experiment):

- Galileo observed that a heavy object and a light object will fall to the ground simultaneously from a particular height.
- All bodies of any mass will fall down simultaneously at the same time from a particular height. That is the acceleration of all free-falling bodies is same. We know that acceleration is the ratio of force and mass.
- The force exerted on a bigger body is greater than the smaller ones in the same force field. This implies acceleration remains constant in the case of all free-falling bodies.

The experiment of inertia:

The property for which a body restrains the force applied to it is called inertia. This is observed by Galileo.

- He noticed that moving a heavy body is very hard than moving a light body.
- No force is exerted on a moving body of constant velocity. We know that force is the product of mass and acceleration. i.e the acceleration is the time rate of change to velocity.
- If velocity remains constant, then acceleration will be zero. i.e, the force exerted on a body moving with constant speed is zero.

The experiment of simple harmonic motion:

- The time period of simple oscillation of a pendulum is constantly in motion.
- The time period of a simple pendulum in simple harmonic motion is constant in time and it depends only on the mass of the pendulum and the string's length.
- This concept helped newton in the formation of his famous law of motion.

The motion of a body on an inclined plane:

- When the force is exerted on a body, then it changes its velocity.
- If a body rolls down on an inclined plane then its velocity will increase due to the effect of the earth's gravitational force exerted on the body.
- This implies that when the force is exerted on a body its velocity will be increase or decrease.

1.2. NEWTON'S LAWS OF MOTION: (1642 – 1726)

There are basically three laws of motion. They are

1.2.1 Newton's first law of motion

1.2.2 Newton's second law of motion

1.2.3 Newton's third law of motion

1.2.1 Newton's First Law of Motion

Statement:

Every object continues in its state of rest or uniform motion in a straight line unless a net external force acts on it to change that state.

(OR)

If a body is at rest or moving at a constant speed in a straight line it will remain same unless it is acted upon by a force.

(OR)

If F_{net} = 0 & a_{net} = 0 then there is no change in state.

$$\therefore \quad \vec{F_1} + \vec{F_2} + \vec{F_3} = 0 \quad \text{------------------- (1)}$$

Example: Inertia & Force

Newton's first law indicates that the state of a body at rest (zero velocity) and a state of uniform velocity are completely equivalent. No external force is needed in order to maintain the uniform motion of a body; it continues without change due to an intrinsic property of the body that we call *inertia*. Because of this property, the first law is often referred to as the *law of inertia*. ***Inertia** is the natural tendency of a body to remain at rest or in uniform motion along a straight line.* Quantitatively, the inertia of a body is measure by its mass. In one sense, Newton made the first law more precise by introducing definitions of quantity of motion and amount of matter which we now call momentum and mass respectively. The momentum of a body is simply proportional to its velocity. The co – efficient of proportionality is a constant for any given body and is called its mass. Denoting mass by m and momentum by **P**

$$P = mv \quad \text{-------------------- (2)}$$

Where V is the velocity of the body. Mathematically, Newton's first law can be expressed in the following way. In the absence of an external force acting on a body

$$P = mv = \text{constant} \qquad \text{-------------- (3)}$$

This is ***law of conservation of momentum***. As per the special theory of relativity, mass is not a constant but varies with velocity.

1.2.2 Newton's Second Law of Motion

Statement:

Net force acting on a body is directly proportional to the rate of change of its linear momentum.

OR

The rate of change of momentum of an object is proportional to the external force and is in the direction of the force.

$$\therefore \qquad \vec{F} \propto \frac{d\vec{P}}{dt} \qquad \text{-------------------- (1)}$$

$$\vec{F} = K\frac{d\vec{P}}{dt}$$

Where, K – is the Proportionality Constant

$$\vec{F} = \frac{d\vec{P}}{dt} \qquad \text{-------------------- (2)}$$

We Know that P = mv

$$\therefore \qquad \vec{F} = \frac{d\,(mv)}{dt} \qquad \text{-------------------- (3)}$$

If mass is constant, then

$$\vec{F} = m\,\frac{dv}{dt}$$

$$\vec{F} = m\,\vec{a} \qquad \text{----------------- (4)}$$

Where a stands for acceleration. Since $V = \frac{dr(t)}{dt}$, the second law may also be written as

$$\vec{F} = m\,\frac{d^2r(t)}{dt^2} \qquad \text{------------------ (5)}$$

Which is often referred to as the equation of motion of the particle. It is a second order differential equation. If the force F is known and the position and velocity of the particle at a particular instant are given, with the help of second law we can find the position and velocity of the particle at any given instant. That is, its path is completely known if accurate values of its co – ordinates and velocity (or momentum P = mv) at a particular instant are known simultaneously. In quantum mechanics, we will be learning that this deterministic model is not applicable to atomic and subatomic particles.

1.2.3 Newton's Third Law of Motion

Statement:

For every action there is always equal and opposite reaction.

(OR)

The mutual action of any two bodies is always equal and oppositely directed along the same straight line.

(OR)

Whenever a body exerts a force on a second body, the second exerts an equal and opposite force on the first.

This law is often paraphrased as to every action there is an equal and opposite reaction. This statement is perfectly valid but it has to be remembered that the action force and the reaction force are acting on different bodies. In a two particle system, the force acting on particle 1 by particle 2, F_{12}, is equal and opposite to the force acting on particle 2 by particle 1, F_{21}. That is,

$$F_{12} = -F_{21} \qquad \text{---------------} (1)$$

Since force is the rate of change of momentum

$$\frac{dP_1}{dt} = -\frac{dP_2}{dt}$$

$$m_1 a_1 = -m_2 a_2 \qquad \text{--------------} (2)$$

$$\frac{m_2}{m_1} = \left|\frac{a_1}{a_2}\right| \qquad \text{---------------} (3)$$

Equation (3) can be used to determine the mass of particles.

Examples

1) A swimmer pushes the water in the backward direction (action) and the water pushes the swimmer in the forward direction (reaction).
   ∴ Forces are equal in magnitude and opposite in direction

2) When a bullet is fired from a gun, the bullet moves forward (action), the gun moves backward (reaction).
3) A block kept on a table exerts a contact force on the table and the table exerts an equal and opposite reaction force on the block.

1.3. FRAMES OF REFERENCE

The most basic concepts for the study of motion are space and time, both of which are assumed to be continuous. To describe the motion of a body, one has to specify its position in space as a function of time. To do this, a co – ordinate system is used as a frame of reference. One convenient co-ordinate system we frequently use is the Cartesian or rectangular co-ordinate system.

Cartesian Co – ordinates (x, y, z)

Cartesian coordinates are commonly used in Physics to describe the position of an object in space. For example, in classical mechanics, the position of a particle can be described using its x, y, and z coordinates. In electromagnetism, the electric and magnetic fields can be described using Cartesian coordinates.

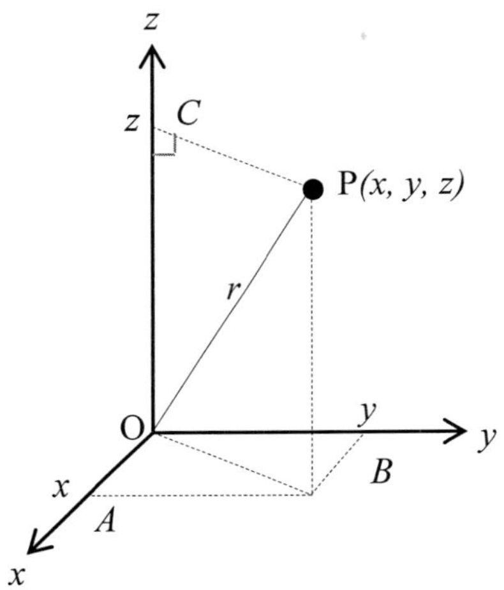

Fig. (1.1): *Cartesian Co – ordinates (x, y, z) of a point P in three dimensions*

The position of a point P in a Cartesian co – ordinate system, as shown in Fig. 1.1. is specified by three co – ordinates (x, y, z) or (x_1, x_2, x_3) or by the position vector **r**. A vector quantity will be denoted by boldface type (as **r**). while the magnitude will be represented by the symbol itself (as **r**). A unit vector in the direction of the vector **r** is denoted by the corresponding letter with a with a circumflex over it (as $\hat{r}$ = $\boldsymbol{r}/r$). In terms of the co – ordinates, the vector and the magnitude of the vector are given by

$$\text{r} = \hat{\imath}x + \hat{\jmath}y + \hat{k}z$$

$$r^2 = x^2 + y^2 + z^2 \qquad \text{------------------------------} \ (1)$$

Where $\hat{\imath}, \hat{\jmath}$ *and* $\hat{k}$ are unit vectors along the rectangular axes x, y and z respectively.

Elementary lengths in the direction of x, y, z: dx, dy, dz
Elementary volume: dx dy dz

Cartesian co – ordinates are convenient in describing the motion of objects in a straight line. However, in certain problems, it is convenient to use non – rectangular co – ordinates.

Plane Polar Co – ordinates *(r, θ)*

In a plane, suppose you have a point O called the origin, and an axis through that point – say the x -axis – called the polar axis. Then the polar coordinates *(r, θ)* describe the point lying a distance of r units away from the origin, at an angle of θ to the x -axis. The value of θ may be given in degrees or radians.

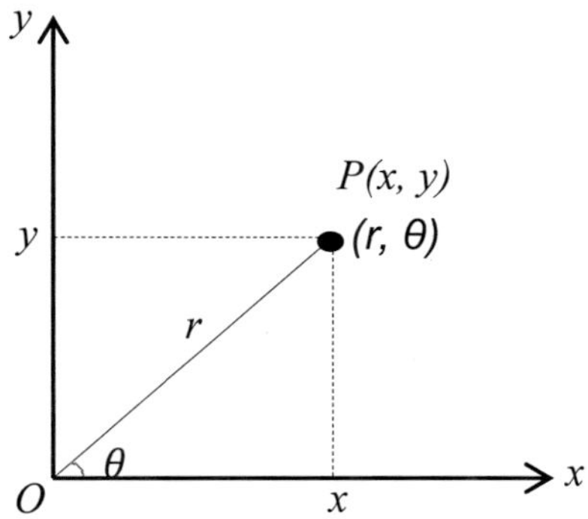

Fig. (1.2): *Plane Polar Co –ordinates (r, θ) of a point P.*

To study the motion of a particle in a plane, the plane co –ordinate system which is shown in Fig. (1.2) is probably the proper choice. The radius vector of the point P in the xy plane is **r**. The line OP makes an angle θ with the x – axis. The position of point P can be described by the co – ordinates (r, θ) called **Plane Polar Co – ordinates**. The rectangular co –ordinates of P are (x, y). The relations connecting (x, y) and (r, θ) can be written from Fig. (1.2) as:

$$x = r\cos\theta \text{ and } y = r\sin\theta \quad \text{---------} (2)$$

From equation (2)

$$r^2 = x^2 + y^2 \text{ and } \theta = \tan^{-1}\left(\frac{y}{x}\right) \text{ --------- (3)}$$

Elementary lengths in the direction of increasing r and θ: dr, rdθ
Elementary area: dr x rdθ = r dr dθ

Velocity components: $\dot{r}$ and $r\dot{\boldsymbol{\theta}}$

Thus, in a two dimensional system *(x, y)* or *(r, θ)* completely specifies the position of a point in a plane.

Cylindrical Co – ordinates (*ρ*, ∅, *z*)

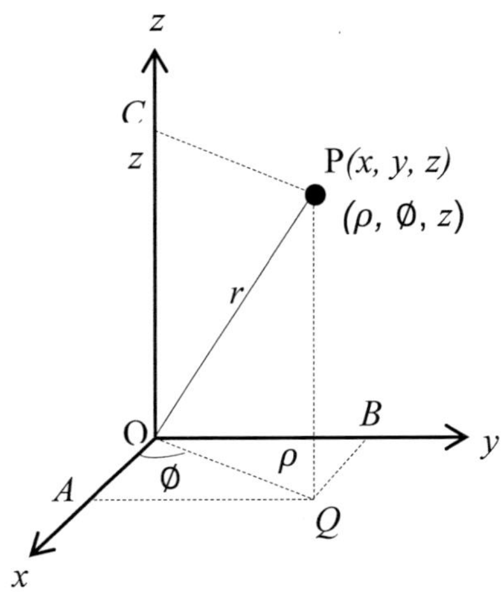

Fig. (1.3): *Cylindrical Co – ordinates (ρ, ∅, z) of a point P in space*

The coordinate surfaces of the cylindrical coordinates $(\rho,\ \emptyset,\ z)$. Consider a point P having a radius vector **r**. Point P can be specified by using a set of Cartesian co –ordinates (x, y, z) or cylindrical co – ordinates $(\rho,\ \emptyset,\ z)$ as shown I Fig. (1.3). The co - ordinate ρ is the projection of the radius vector **r** on the *xy* – plane. The two sets of co – ordinates are related by the relations:

$$x = \rho \cos\emptyset \qquad \rho^2 = x^2 + y^2$$

$$y = \rho \sin\emptyset \qquad \emptyset = \tan^{-1}\left(\frac{y}{x}\right) \text{ ------------- (4)}$$

Elementary lengths: *dp, ρd∅, dz*
Elementary Volume: *ρ dρ d∅ dz*
Velocity components: $\dot{\rho}$, $\rho\dot{\emptyset}$, $\dot{z}$

Spherical Polar Co – ordinates *(r, θ,* ∅)

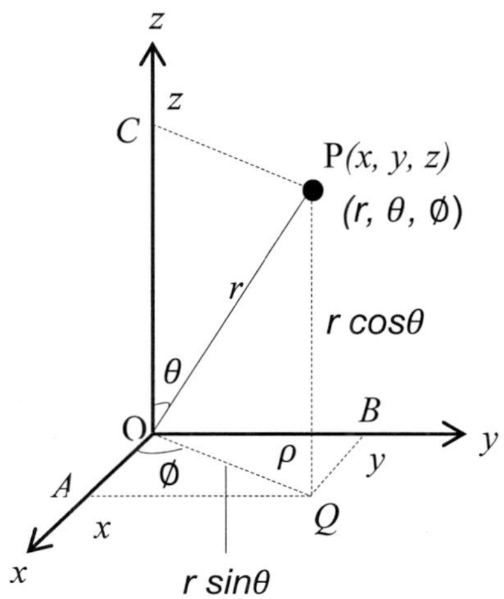

Fig. (1.4): *Spherical Polar Co – ordinates (r, θ, ∅) of a point P in space*

Figure (1.4) defines the spherical polar co – ordinates of a point P having a radius vector **r**. The Cartesian co – ordinates of P are *(x, y, z)*. The co –ordinate θ is the angle that the radius vector r makes with the z – axis and ∅ is the angle that the projection of the position vector into the xy – plane makes with x – axis. From Fig. (1.4).

The two sets of co –ordinates are related by the relations:

$x = r \sin\theta \cos\emptyset$
$y = r \sin\theta \sin\emptyset$
$z = r \cos\theta$ ---------------------- (5)

Elementary lengths: $dr,\ rd\theta\ r\sin\theta\ d\emptyset$
Elementary volume: $dr \times rd\theta \times r\sin\theta\ d\emptyset = r^2 \sin\theta\ d\theta\ d\emptyset\ dr$
Elementary components: $\dot{r},\ r\dot{\theta},\ \ r\sin\theta\ \dot{\emptyset}$

1.4. INERTIAL AND NON – INERTIAL FRAME OF REFERENCE

Frame of reference is a coordinate system relative to which the position & motion of a body may be described the simple frame of reference is the Cartesian coordinate system in which every point of space may be described by 3 numbers *(x, y, z)* be the 3 coordinates of that point, if $\vec{r}$ is position vector of point P relative to origin O of a Cartesian frame of reference then $\vec{r} = x\hat{\imath} + y\hat{\jmath} + z\hat{k}$. In order to specify the exact time of knowing the position of the given particle, we must assign one more co – ordinate t, the time co – ordinate. Such a reference frame is called space – time reference frame & is consists of four co –ordinates (x, y, z, t)

Types of Frame of Reference

There are 2 types of frame of reference

1) Inertial frame of reference
2) Non – Inertial frame of reference

1) Inertial frame of reference

A frame of reference in which Newton's laws of motion holds good is known as inertial frame of reference. It means a frame of reference which either at rest or moves with uniform velocity is an inertial frame of reference.

From Newton's 2nd Law of motion

$$\vec{F} = m\,\vec{a}$$

If $\vec{F} = 0$ Then $m\,\vec{a} = 0$

As $m \neq 0$

So $\vec{a} = 0$

Thus a inertial frame of reference is also known as unaccelerated or non – accelerating frame of reference.

2) Non – Inertial frame of reference

A frame of reference in which Newton's laws of motion are not valid is known as non – inertial frame of reference.

Thus a non – inertial frame of reference is an accelerated frame of reference.

The accelerations equation (3) (In Newton's 3rd law of motion) can be measured experimentally. Hence, equation (3) can be used to determine the mass of a particle by selecting m_1 as unit mass. The mass of a body determined in this way is termed as its inertial mass because it characterizes the inertial properties of bodies. Mass can also be defined on the basis of gravitational properties is called the ***gravitational mass***. Naturally a question arises; Is the inertial mass of a body equal to its gravitational mass? Recently it was established that these masses are equal to within a few parts in 10^{12}. This equivalence of the inertial and gravitational masses of a body is the ***principle of equivalence*** postulated by Einstein in general relativity.

1.5. MECHANICS OF A PARTICLE

In this section, we shall discuss mainly the conservation laws for a particle in motion in Newtonian formalism.

1.5.1. Mechanics of a Single Particle

(OR)

Newtonian Mechanism of a Single Particle (or) Point Particle

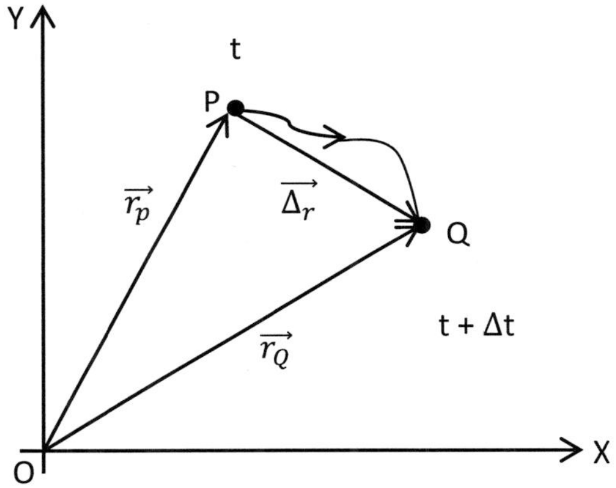

Fig. (1.5): *Mechanism of Single Particle*

Consider two dimensional frames of reference in a plane. Let P be the point on the plane and $\overrightarrow{r_p}$ & $\overrightarrow{r_Q}$ be the position vectors of Points P & Q from Origin 'O'. Let $\Delta\vec{r}$ be the displacement vector.

Now Velocity of the particle is defines as

$$V = \lim_{\Delta t \to 0} \frac{\Delta\vec{r}}{\Delta t}$$

$$\therefore \qquad V = \frac{d\vec{r}}{dt} \qquad \text{---------------- (1)}$$

Acceleration of the particle is given by

$$\vec{a} = \frac{d\,(v)}{dt}$$

$$\vec{a} = \frac{d}{dt}\left(\frac{d\vec{r}}{dt}\right)$$

$$\vec{a} = \frac{d^2 r}{dt^2} \qquad \text{-------------------- (2)}$$

If we apply some external force on particle 'P'. Now we observe how particle 'P' reacts with external force.

According to Newton's Second law of Motion

$$\vec{F} = \frac{d\vec{P}}{dt} \quad \text{------------------ (3)}$$

Or $\vec{F} = \dot{\vec{P}}$

We know that Linear Momentum of the particle is

$$\vec{P} = \text{m}\,\vec{V}$$

Put this $\vec{P} = \text{m}\,\vec{V}$ Value in Equation (3)

∴ Equation (3) becomes $\vec{F} = \frac{d\,(m\vec{v})}{dt}$

If mass is constant

$$\vec{F} = \text{m}\,\frac{d\,(\vec{v})}{dt}$$

$$\vec{F} = \text{m}\,\vec{a} \quad \text{------------------ (4)}$$

$$\vec{F} = \text{m}\,\frac{d^2r}{dt^2} \quad \text{------------------ (5)}$$

$$\vec{F} = \ddot{\vec{r}}$$

Law of Conservation of Linear Momentum of a particle

According to the law of conservation of linear momentum for a point particle

From equation (3) $\vec{F} = \frac{d\vec{P}}{dt}$

Consider if no any force is acting on the particle i.e $\vec{F}_{net} = 0$

Then $\frac{d\vec{P}}{dt} = 0$

Integrate on both sides

$$\int \frac{d\vec{P}}{dt} = \int 0$$

$$\vec{P} = 0 \quad \text{------------------ (6)}$$

∴ $\vec{P}$ = Constant

If there is no any force is acting on a particle or the vector sum of all the forces is zero then the linear momentum of the particle is conserved.

Law of Conservation of Angular Momentum of a particle
Let L be the angular momentum of the particle 'P'.

∴ $L = \vec{r} \times \vec{P}$ ------------------- (7)

Where,
$\vec{r}$ – is the position vector
$\vec{P}$ – is the linear momentum of the particle at time 't'.

The rate of change of angular momentum 'L' is defined as Torque.

$$\therefore \quad \tau = \frac{dL}{dt}$$

$$\tau = \frac{d\,(\vec{r} \times \vec{P}\,)}{dt}$$

$$\tau = \vec{r}\,\frac{d\vec{P}}{dt} + \vec{P}\,\frac{d\vec{r}}{dt}$$

$$\tau = \vec{r}\,\frac{d\vec{P}}{dt} + m\,\vec{V}.\,\vec{V} \qquad \begin{pmatrix} We\ Know\ that \\ \vec{P} = m\,\vec{V} \\ \frac{d\vec{r}}{dt} = \vec{V} \end{pmatrix}$$

$$\tau = \vec{r}\,\frac{d\vec{P}}{dt} + 0 \qquad \begin{pmatrix} We\ Know\ that \\ \vec{V}.\vec{V} = 0 \end{pmatrix}$$

$$\tau = \vec{r}.\,\vec{F} \text{ -------------------- (8)}$$

Thus Torque is equal to the momentum of the given point.
If Torque acting on the particle is zero. Then we have

$$\tau = \frac{dL}{dt}$$

$$0 = \frac{dL}{dt}$$

Integrate on both sides

$$\int 0 = \int \frac{dL}{dt}$$

$$0 = L$$

∴ $L = 0$ ----------------------- (9)

$L = \text{Constant}$

Thus, If no force acting on the particle it's angular momentum is constant or conserved

Law of Conservation of Energy of a particle

Statement:

"It states that if forces acting on a particle are conservative then total energy (P.E + K.E) of the particle is constant or conserved."

OR

"Under conservative forces, the total mechanical energy of the system remains conserved".

Proof:

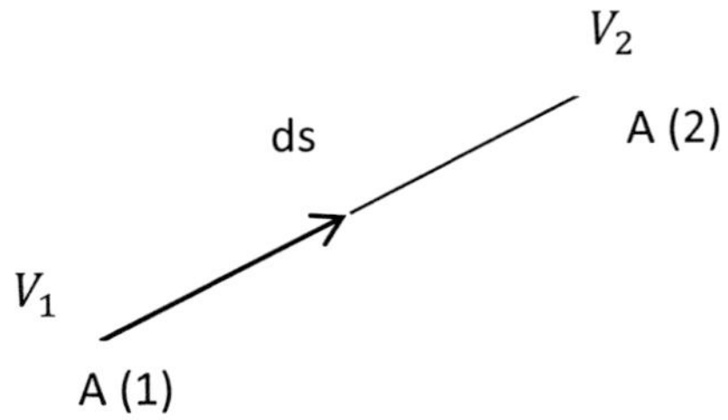

Let Force $\vec{F}$ is applied on a particle of mass 'm' and small amount of work is done by Force $\vec{F}$ in displacing the particle through small displacing is given by

$$dW = \vec{F} \cdot d\vec{S} \quad \text{------------------ (10)}$$

The total work done by force is displacing the particle from A to B is given by

Integrating equation (10) on both sides from 1 to 2, we get

$$\int_1^2 dW = \int_1^2 \vec{F} \cdot d\vec{S} \quad \text{------------------ (11)}$$

$$W_{12} = \int_1^2 \vec{F} \cdot d\vec{S}$$

$$W_{12} = \int_1^2 m\vec{a} \cdot d\vec{S} \qquad \begin{pmatrix} We\ Know\ that \\ \vec{F} = m\vec{a} \end{pmatrix}$$

$$W_{12} = \text{m} \int_1^2 \vec{a} \cdot d\vec{S}$$

$$W_{12} = \text{m} \int_1^2 \frac{d\vec{\vec{V}}}{dt} \cdot d\vec{S}$$

$$W_{12} = \text{m} \int_1^2 \frac{d\vec{\vec{V}}}{dt} \cdot d\vec{S} \cdot \frac{dt}{dt} \quad (Multiply\ and\ Divide\ with\ dt)$$

$$W_{12} = \text{m} \int_1^2 \frac{d\vec{\vec{V}}}{dt} \cdot \frac{d\vec{S}}{dt} \cdot \text{dt}$$

$$W_{12} = \text{m} \int_1^2 \frac{d\vec{\vec{V}}}{dt} \cdot \vec{V} \cdot \text{dt} \qquad \begin{pmatrix} We\ Know\ that \\ \frac{d\vec{S}}{dt} = \vec{V} \end{pmatrix}$$

$$W_{12} = \text{m} \int_1^2 d\vec{V} \,.\vec{V}$$

$$W_{12} = \text{m} \int_1^2 \vec{V} \,.d\vec{V}$$

Now integrate $\vec{V}$ with respect to $d\vec{V}$, we get

$$W_{12} = \text{m} \left[\frac{V^2}{2}\right]_1^2$$

$$W_{12} = \frac{m}{2} [V^2]_1^2$$

$$W_{12} = \frac{m}{2} \left[{V_2}^2 - {V_1}^2\right]$$

$$W_{12} = \frac{1}{2}\, m\, {V_2}^2 - \frac{1}{2}\, m\, {V_1}^2$$

$$W_{12} = T_2 - T_1 \qquad \text{---------------------- (12)}$$

Where T_1 & T_2 are the K.E of the particle in position 1 & 2 respectively.

If $T_1 > T_2$ Then $W_{12} < 0$

i.e. Work is done by the particle against the force and its K.E as decreased.
If $T_2 > T_1$ Then $W_{12} > 0$
i.e. Work is done by the force on the particle & the K.E of the particle is increased.

$$\vec{F} = - \frac{dv}{ds}$$

$$\vec{F} \,.\, d\vec{s} = - \text{dv}$$

Integrating on both sides from 1 to 2, we get

$$\int_1^2 \vec{F} \,.d\vec{s} = - \int_1^2 dv$$

$$W_{12} = - [V]_1^2$$

$$W_{12} = - [V_2 - V_1]$$

$$W_{12} = V_1 - V_2 \qquad \text{------------------------- (13)}$$

Equating equations (12) & (13), we get

$$T_2 - T_1 = V_1 - V_2$$

$$T_2 + V_2 = T_1 + T_1$$

$$E_1 = E_2$$

Generally E_1 = E_2 = E_3 = E_4 = E = Constant or Conserved

Thus, the sum of potential & Kinetic Energies of a particle at every point in a conservative force field is constant.

1.5.2. Mechanics of a System of Particles

(OR)

Mechanics of Many Particles

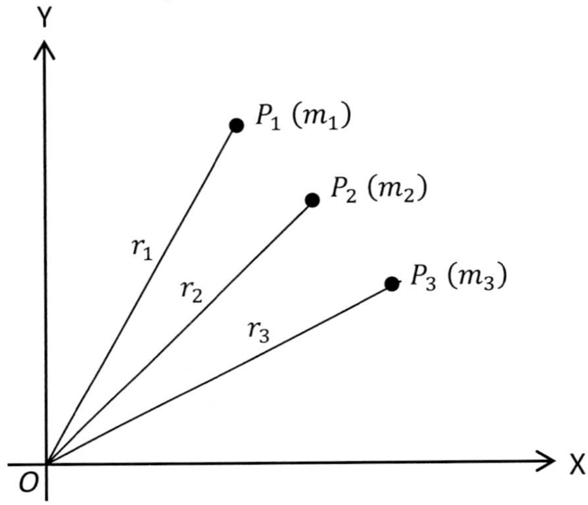

Fig. (1.6): *Mechanics of a System of Particles*

Consider a system of nth particles with Positions P_1, P_2, P_3 ---------- P_n and Position vectors r_1, r_2 , r_3 -------- r_n having the masses m_1 , m_2 , m_3 -------- m_i from the origin. Let $F_{external}$ be the external force acting on the system. Consider i^{th} particle of the system having the mass m_i .

Let us define a vector $\vec{R}$ such that it is an average of $\vec{r_i}$ weighted with their respective masses.

$$\vec{R} = \frac{m_1\,\vec{r_1} + m_2\,\vec{r_2} + --------- + m_n\,\vec{r_n}}{m_1 + m_2 + -------- + m_n}$$

$$\vec{R} = \frac{\sum_i m_i\,\vec{r_i}}{\sum_i m_i}$$

$$\vec{R} = \frac{\sum_i m_i\,\vec{r_i}}{M}$$

Where M – is the mass of the whole system

$$\therefore \qquad M\,\vec{R} = \sum_i m_i\,\vec{r_i} \qquad ------------------------ (1)$$

Let F_{ij} be the internal force acting on the ith particle and j^{th} particle as a result of it's interaction with j^{th} particle.

Then, the total internal force acting on the ith particle is given by

$$F_i^{internal} = \sum_j F_{ij}^{internal} \qquad (i \neq j)$$

Then the equation of motion for the ith particle or Newton's Second law of motion can be written as

$$\vec{\dot{P_i}} = \vec{F_i}^{ext} + \sum_j \vec{F_{ij}}^{int} \qquad \text{------------------- (2)}$$

Where,

$\vec{F_i}^{ext}$ – is the external force on the system.

$\vec{F_{ij}}^{int}$ – is the internal force on i^{th} particle due to j^{th} particle.

We assume that $\sum_j \vec{F_{ij}}^{int}$ obeys Newton's 3rd law of motion. So we can write

$$\vec{F_{ij}}^{int} = -\vec{F_{ji}}^{int} \qquad \text{---------------- (3)}$$

And $\vec{F_{ij}}^{int} = 0$

Now, equation (2) is summing over ' i ' to give the equation of motion of the system.

$$\sum_i \vec{\dot{P_i}} = \sum_i \vec{F_i}^{ext} + \sum_i \sum_j \vec{F_{ij}}^{int} \qquad \text{----------------- (4)}$$

$$\sum_i \vec{\dot{P_i}} = \sum_i \vec{F_i}^{ext} + \sum_{ij} \vec{F_{ij}}^{int}$$

Here $\sum_{ij} \vec{F_{ij}}^{int} = 0$ & $\sum_i \vec{F_i}^{ext} = \vec{F}^{ext}$

$$\sum_i \frac{d\vec{P}_i}{dt} = \vec{F}^{ext} + 0$$

$$\sum_i \frac{d\,(m_i\vec{v}_i)}{dt} = \vec{F}^{ext}$$

$$\sum_i \frac{d}{dt}\left(\frac{dr_i}{dt}\right) m_i = \vec{F}^{ext}$$

$$\sum_i \frac{d^2 r_i}{dt^2} m_i = \vec{F}^{ext}$$

$$\frac{d^2}{dt^2} \sum_i (m_i \vec{r}_i) = \vec{F}^{ext}$$

$$\frac{d^2}{dt^2}\left(M\vec{R}\right) = \vec{F}^{ext}$$

$$M\,\frac{d^2\vec{R}}{dt^2} = \vec{F}^{ext} \qquad \text{-------------------------- (5)}$$

Here $\vec{R}$ defines a point known as centre of mass or centre of gravity.

Law of Conservation of Linear Momentum for a System of Particles

The equation of motion (Newton's 2nd law of motion) for a system of particles is given by

$$\vec{F}^{ext} = M\,\frac{d^2\vec{R}}{dt^2}$$

Where,

$\vec{F}^{ext}$ – is vector sum of all the external forces

M – is mass of whole system

$\vec{R}$ – is position vector of centre of mass

$$\vec{F}^{ext} = M\,\frac{d}{dt}\left(\frac{d\vec{R}}{dt}\right)$$

$$\vec{F}^{ext} = M\,\frac{d\vec{V}_{cm}}{dt}$$

Where,

$\vec{V}_{cm}$ – is the velocity of centre of mass

As we have assumed the mass of the system is constant.

$$\vec{F}^{ext} = \frac{dM\vec{V}_{cm}}{dt}$$

$$\vec{F}^{ext} = \frac{d\vec{P}}{dt}$$

Where,

$\vec{P}$ – is linear momentum of centre of mass of the system of particles.

If there is no any external force is acting on the system or the vector sum of all the external force is zero. i.e., If $\vec{F}^{ext}$ = 0 Then

$$\frac{d\vec{P}}{dt} = 0 \qquad \text{--------------------- (6)}$$

$\therefore$ $\vec{P}$ = Constant or Conserved

Theorem: ANGULAR MOMENTUM

Prove that the angular momentum of the system about a fixed point is equal to the sum of the angular momentum of the total mass concentrated at the centre of mass about that point and angular momentum of an system about it's centre of mass.

Proof:

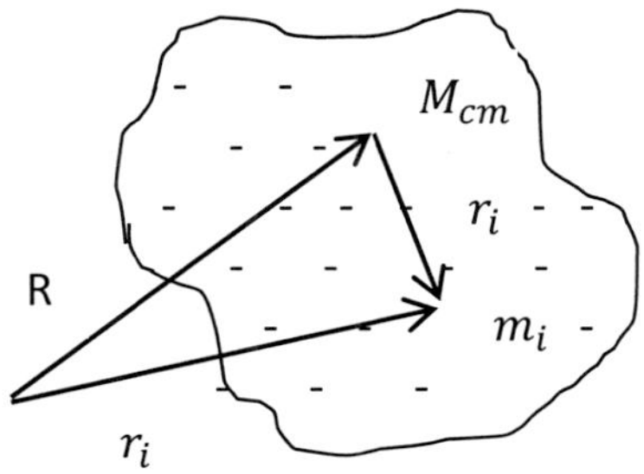

Fig. (1.7): *System of Particles with total mass 'M'& Centre of mass at R*

Let m_i & r_i denote mass & the position vector of the ith particle with reference to point 'O' as shown n figure (1.7).

Let r_i be the position vector of the ith particle with reference to the centre of mass of the system, then

$$r_i = R + r_i{}' \quad \text{------------------------- (a)}$$

The corresponding equation relating to the velocities can be written as

$$\dot{r}_i = \dot{R} + \dot{r}_i{}'$$
$$V_i = V + V_i{}' \quad \text{----------------------- (b)}$$

Where

V – is the velocity of centre of mass of the system

$V\,'$ - is the velocity of ith particle w.r.to. Centre of mass of the system

The total angular momentum is given by

$$L = \sum_i r_i \, X \, P_i$$

$$L = \sum_i (R + r_i{}') \, X \, m_i \left(\dot{R} + \dot{r}_i{}'\right)$$

$$L = \sum_i m_i \left(R \, X \, \dot{R}\right) + \left(R \, X \, \sum_i m_i \dot{r}_i{}'\right) + \left(\sum_i m_i r_i{}' \, X \, \dot{R}\right) + \sum_i (m_i r_i{}' \, X \, \dot{r}_i{}')$$

The quantity $\sum_i m_i r_i{}^{\prime}$ vanishes (i.e. $\sum_i m_i r_i{}^{\prime} = 0$) as it defines the radius vector of the centre of mass in the co-ordinate system in which origin is the centre of mass. The 2nd term can be written as

$$\sum_i m_i \dot{r}_i{}^{\prime} = \frac{d}{dt}(\sum_i m_i r_i{}^{\prime})$$

$$= 0$$

Hence, the total angular momentum

$$L = \sum_i m_i \left(R\, X\, \dot{R}\right) + \sum_i (m_i r_i{}^{\prime}\ X \dot{r}_i{}^{\prime})$$

$$L = \left(R\, X\, \dot{R}\right) \sum_i m_i + \sum_i r_i{}^{\prime}\, X\, m_i \dot{r}_i{}^{\prime}$$

$$L = R \times \dot{R}\, M + \sum_i r_i{}^{\prime}\, X\, P_i{}^{\prime}$$

$$L = R \times P + L^{\prime} \qquad \text{------------------------} \ (c)$$

Where

P = M$\dot{R}$ represents the linear momentum of the centre of mass and

R x P represents the angular momentum of the total mass concentrated at the centre of mass about point 'O'.

The term $L^{\prime} = \sum_i r_i{}^{\prime}\, X\, P_i{}^{\prime}$ represents the angular momentum of the system about its centre.

Hence, it is proved

Law of Conservation of Angular Momentum for a System of Particles

The angular momentum of the ith particle is given by

$$\vec{l}_i = \vec{r}_i\, X\, \vec{P}_i$$

For angular momentum of the whole system we have to sum it over i = 1 to n

$$\sum_{i=1}^{n} \vec{l}_i = \sum_{i=1}^{n} (\vec{r}_i\, X\, \vec{P}_i)$$

$$\vec{L} = \sum_{i=1}^{n} (\vec{r}_i\, X\, \vec{P}_i)$$

Differentiating with respect to time, we get

$$\frac{d\vec{L}}{dt} = \sum_i \frac{d}{dt}(\vec{r}_i\, X\, \vec{P}_i)$$

$$\frac{d\vec{L}}{dt} = \sum_i \left(\vec{r}_i \, X \, \frac{d\vec{P}_i}{dt} + \frac{d\,\vec{r}_i}{dt} \, X \, \vec{P}_i \right)$$

$$\frac{d\vec{L}}{dt} = \sum_i \left(\vec{r}_i \, X \, \frac{d\vec{P}_i}{dt} + \vec{V}_i \, X \, m_i \vec{V}_i \right)$$

$$\frac{d\vec{L}}{dt} = \sum_i \left(\vec{r}_i \, X \, \frac{d\vec{P}_i}{dt} + 0 \right) \qquad \begin{pmatrix} We\ Know\ that \\ \vec{V}.\vec{V} = 0 \end{pmatrix}$$

$$\frac{d\vec{L}}{dt} = \sum_i \left(\vec{r}_i \, X \, \frac{d\vec{P}_i}{dt} \right)$$

$$\frac{d\vec{L}}{dt} = \sum_i \left(\vec{r}_i \, X \, \vec{F}_i \right) \qquad \text{----------------------- (7)}$$

From the equation of motion for a system of particles

$$\sum_i \frac{d\vec{P}_i}{dt} = \sum_i \vec{F}_i^{\,ext} + \sum_{ij} \vec{F}_{ij}^{\,int}$$

Taking cross product with $\vec{r}_i$ on both sides

$$\sum_i \left(\vec{r}_i \, X \frac{d\vec{P}_i}{dt} \right) = \sum_i \left(\vec{r}_i \, X \, \vec{F}_i^{\,ext} \right) + \sum_{ij} \left(\vec{r}_i \, X \, \vec{F}_{ij}^{\,int} \right)$$

$$\sum_i \left(\vec{r}_i \, X \, \vec{F}_i \right) = \sum_i \left(\tau_i^{\,ext} \right) + \sum_{ij} \left(\vec{r}_i \, X \, \vec{F}_{ij}^{\,int} \right) \qquad \text{------ (8)}$$

From equation (7) we know that $\frac{d\vec{L}}{dt} = \sum_i \left(\vec{r}_i \, X \, \vec{F}_i \right)$, put this value in equation (8), we get

$$\frac{d\vec{L}}{dt} = \tau^{ext} + \sum_{ij} \left(\vec{r}_i \, X \, \vec{F}_{ij}^{\,int} \right) \qquad \text{--------- (9)}$$

Now taking Second term of equation (9)

$$\sum_{ij} \left(\vec{r}_i \, X \, \vec{F}_{ij}^{\,int} \right) = \sum_{ij} \left(\vec{r}_j \, X \, \vec{F}_{ji}^{\,int} \right) \begin{pmatrix} Here\ i\ j\ are \\ interchangible \end{pmatrix} \text{-------------- (10)}$$

Again taking Second term of equation (9)

$$\sum_{ij} \left(\vec{r}_i \, X \, \vec{F}_{ij}^{\,int} \right) = \frac{1}{2} \sum_{ij} 2 \left(\vec{r}_i \, X \, \vec{F}_{ij}^{\,int} \right) \qquad (Multiply\ \&Divide\ with\ 2)$$

$$= \frac{1}{2} \sum_{ij} \left(\vec{r}_i \, X \, \vec{F}_{ij}^{\,int} + \vec{r}_i \, X \, \vec{F}_{ij}^{\,int} \right)$$

$$= \frac{1}{2}\sum_{ij}\left(\vec{r}_i \, X \, \vec{F}_{ij}^{\;int} + \vec{r}_j \, X \, \vec{F}_{ji}^{\;int}\right) \quad (From\ eq'n\ 10) \text{ ----- } (11)$$

We know that the internal Forces obey Newton's 3rd law of motion

$$\vec{F}_{ij}^{\;int} = -\,\vec{F}_{ji}^{\;int} \quad \text{----------------- } (12)$$

Put equation (12) value in equation (11), we get

$$\sum_{ij}\left(\vec{r}_i \, X \, \vec{F}_{ij}^{\;int}\right) = \frac{1}{2}\sum_{ij}\left(\vec{r}_i \, X \, \vec{F}_{ij}^{\;int} + \vec{r}_j \, X \left(-\,\vec{F}_{ij}^{\;int}\right)\right)$$

$$\sum_{ij}\left(\vec{r}_i \, X \, \vec{F}_{ij}^{\;int}\right) = \frac{1}{2}\sum_{ij}\left(\vec{r}_i \, X \, \vec{F}_{ij}^{\;int} - \vec{r}_j \, X \, \vec{F}_{ij}^{\;int}\right)$$

$$\sum_{ij}\left(\vec{r}_i \, X \, \vec{F}_{ij}^{\;int}\right) = \frac{1}{2}\sum_{ij}(\vec{r}_i - \vec{r}_j) \, X \, \vec{F}_{ij}^{\;int} \quad \text{---------------- } (13)$$

$$\sum_{ij}\left(\vec{r}_i \, X \, \vec{F}_{ij}^{\;int}\right) = \frac{1}{2}\sum_{ij}(\vec{r}_i - \vec{r}_j) \, X \, \vec{F}_{ij}^{\;int} \quad \begin{pmatrix} Using\ triangle\ of \\ vector\ addition \\ \vec{r}_j + \vec{r}_{ij} = \vec{r}_i \\ \vec{r}_{ij} = \vec{r}_i - \vec{r}_j \end{pmatrix}$$

$$\sum_{ij}\left(\vec{r}_i \, X \, \vec{F}_{ij}^{\;int}\right) = \frac{1}{2}\sum_{ij}\vec{r}_{ij} \, X \, \vec{F}_{ij}^{\;int} \quad \text{------------------ } (14)$$

Put equation (14) value in equation (9), we get

$$\frac{d\vec{L}}{dt} = \tau^{ext} + \frac{1}{2}\sum_{ij}\vec{r}_{ij} \, X \, \vec{F}_{ij}^{\;int} \quad \text{------------------- } (15)$$

Assume that the internal forces are central forces. Then the force $\vec{F}_{ij}$ will be either parallel or antiparallel to $\vec{r}_{ij}$ i.e., θ will be either o^0 or 180^0. In both case

$$\vec{r}_{ij} \, X \, \vec{F}_{ij}^{\;int} = 0$$

Put this value in equation (15), we get

$$\frac{d\vec{L}}{dt} = \tau^{ext} + \frac{1}{2}\sum_{ij} 0$$

$$\frac{d\vec{L}}{dt} = \tau^{ext} + 0$$

$$\frac{d\vec{L}}{dt} = \tau^{ext}$$

If Torque acting on a system is zero, then the angular momentum of the system is constant or conserved.

i.e., If $\tau^{ext} = 0$

Then $\frac{d\vec{L}}{dt} = 0$

On Integrating on both sides, we get

$$\int \frac{d\vec{L}}{dt} = \int 0$$

$$\vec{L} = 0 \quad \text{---------------------- (16)}$$

Or $\vec{L}$ = Constant or Conserved

Law of Conservation of Energy for a System of Particles

The equation of motion for i^{th} particle is given by

$$\vec{F_i} = \vec{F_i}^{ext} + \sum_j \vec{F}_{ij}$$

The work from configuration 1 to 2 for the whole system is given by

$$W_{12} = \sum_i \int_1^2 \vec{F_i} d\vec{r_i} = \sum_i \int_1^2 \vec{F_i}^{ext} d\vec{r_i} + \sum_{ij} \int_1^2 \vec{F}_{ij} d\vec{r_i} \quad (i \neq j) \quad \text{------ (17)}$$

Consider the Left term

$$\sum_i \int_1^2 \vec{F_i} d\vec{r_i} = \sum_i \int_1^2 m_i \dot{\vec{V_i}} d\vec{r_i} \quad \text{---------- (18)}$$

$\left(\begin{array}{c} We\ know\ that \\ \vec{F} = m\vec{a} \\ \vec{F} = m\,\frac{d^2\vec{r}}{dt^2} \\ \vec{F} = m\dot{\vec{V}} \end{array}\right)$

Here $\dot{\vec{V_i}} = \frac{d\vec{V_i}}{dt}$

$$\dot{\vec{V_i}} = \frac{d\vec{r_i}}{d\vec{r_i}} \times \frac{d\vec{V_i}}{dt} \qquad (Multiply\ and\ Divide\ with\ d\vec{r_i})$$

$$\dot{\vec{V_i}} = \frac{d\vec{r_i}}{dt} \times \frac{d\vec{V_i}}{d\vec{r_i}}$$

$$\dot{\vec{V_i}} = \vec{V_i} \times \frac{d\vec{V_i}}{d\vec{r_i}} \quad \text{-------------- (19)}$$

Put equation (19) value in equation (18), we get

$$\sum_i \int_1^2 \vec{F_i} d\vec{r_i} = \sum_i \int_1^2 m_i \vec{V_i} \times \frac{d\vec{V_i}}{d\vec{r_i}}\, d\vec{r_i}$$

$$= \sum_i m_i \int_1^2 \vec{V}_i \, d\vec{V}_i$$

$$= \sum_i m_i \left[\frac{\vec{V}_i^{\,2}}{2}\right]_1^2$$

$$= \sum_i \left(\frac{1}{2} m_i \vec{V}_i^{\,2}\right)_2 - \sum_i \left(\frac{1}{2} m_i \vec{V}_i^{\,2}\right)_1$$

$$\sum_i \int_1^2 \vec{F}_i d\vec{r}_i = T_2 - T_1 \qquad \text{---------------- (20)}$$

This is the difference in K. E of the whole System at two configurations.
Now consider the first term on right hand side

i.e. $$\sum_i \int_1^2 \vec{F}_i^{\,ext} \, d\vec{r}_i \qquad \text{------------------ (21)}$$

If $\vec{F}_i^{\,ext}$ can be expressed as negative gradient of some scalar potential function (Conservative force)

$$\vec{F}_i^{\,ext} = - \frac{d\vec{V}_i}{d\vec{r}_i} \qquad \text{------------------ (22)}$$

Put Value of $\vec{F}_i^{\,ext}$ in equation (21), we get

$$\sum_i \int_1^2 \vec{F}_i^{\,ext} \, d\vec{r}_i = \sum_i \int_1^2 - \frac{d\vec{V}_i}{d\vec{r}_i} \, d\vec{r}_i$$

$$= \sum_i [-\vec{V}_i]_1^2$$

$$= (\sum_i - \vec{V}_i)_2 - (\sum_i - \vec{V}_i)_1$$

$$\sum_i \int_1^2 \vec{F}_i^{\,ext} \, d\vec{r}_i = V_1 - V_2 \qquad \text{----------------- (23)}$$

From equation (17)

$$\sum_i \int_1^2 \vec{F}_i d\vec{r}_i = \sum_i \int_1^2 \vec{F}_i^{\,ext} \, d\vec{r}_i + \sum_{ij} \int_1^2 \vec{F}_{ij} d\vec{r}_i \ (i \neq j)$$

Put the values of equation (20) & (21) in equation (17), we get

$$T_2 - T_1 = V_1 - V_2 + 0$$

$$T_2 - T_1 = V_1 - V_2$$

$$T_1 + V_1 = T_2 + V_2$$

$$E_1 = E_2 \qquad \text{--------------------} \ (24)$$

$\therefore \ E_1 = E_2 = E_3 = E_4 = E_N$ = Constant

Equation (24) states that the total energy of the system is constant or conserved.

This is true only when all the forces (internal & external) forces are derivable from the P.E functions that do not depend on time.

A system in which all the forces acting on it are derivable from the P.E function is called a conservative system.

1.6. APPLICTIONS OF NEWTONIAN MECHANICS

The applications of Newtonian mechanics are based on three laws of Newtonian mechanics principles. Each law has equal interpretation and is equally related to each other. These fundamental laws affect the physical properties of materials as well as their internal properties.

1) The first law of Newtonian mechanics states that only external force application will cause the motion or stop the object. By this application, we can observe in our routine life that every physical object needs an external force to change its initial state. For example stopping balls, Moving cars etc. This law is also called the inertia law of Newtonian mechanics.
   For example: stopping balls, Moving cars, etc.

2) The second law of Newtonian mechanics states that the applied force depends on the product of mass and acceleration. It means higher acceleration gain to the low mass object or high applied force. In real life, when an object gets high speed, its weight reduces.

3) The third law of Newtonian mechanics states that the action of a body will cause an equal and opposite reaction. This is mostly applied law in Newtonian mechanics as action – reaction law. Rocket propulsion uses this law to get sufficient thrust for life – off.

4) Newtonian mechanics involves constraint forces.

5) Newtonian mechanics is used for vectors.

6) This theory is applied to non – conservative forces as well.

7) Newtonian mechanics is related to our routine life as the application of force to do something lifting objects, throwing of disc, lifting of rockets works obey different laws of Newtonian mechanics.

8) Newtonian mechanics has been used to calculate the orbits of planets and astronauts in the heavens and to calculate the motions of objects on the earth.

9) Newtonian mechanics is used extensively in engineering to design and analyze structures, machines, and systems.

10) Spacecraft operate at very high velocities compared to velocities we are familiar with in transportation and ballistics here on our planet. Since spacecraft velocities do not approach a significant fraction of the speed of light, Newtonian physics serves well for operating and navigating throughout the solar system.

1.7. EQUATION OF MOTION FOR A PARTICLE

In general the force acting on a particle may depend on its position, velocity and time; hence equation of motion of a particle is given by

$$F = F\,(r, \dot{r}, t)$$

$$F\ddot{r} = F\,(r, \dot{r}, t) \qquad \text{----------------} \ (1)$$

$$m\,\frac{d^2r}{dt^2} = F\,(r, \dot{r}, t) \qquad \text{---------------} \ (2)$$

This is a 2nd ordered differential equation, we have to solve this by integrating twice we get, Trajectory of the particle there will be 2 constants of integration introduced and hence will be determined by knowing initial conditions of the particle.

Integrating equation (1) w.r.to. Time, we get

$$m \int_{t_o}^{t} \ddot{r} dt^{\prime} = \int_{t_o}^{t} F dt^{\prime}$$

$$m [\dot{r}]_{t_o}^{t} = \int_{t_o}^{t} F dt^{\prime}$$

$$[m \dot{r}]_{t} - [m \dot{r}]_{t_o} = \int_{t_o}^{t} F dt^{\prime}$$

$$mv - mV_o = \int_{t_o}^{t} F dt^{\prime} \qquad \begin{pmatrix} We\ know\ that \\ r^{\prime} = V \\ r_o^{\prime} = V_o \end{pmatrix}$$

$$m (v - v_0) = \int_{t_o}^{t} F dt^{\prime} \quad \text{------------------ (3)}$$

Where we have initially at t = t_0 ∴ Velocity $\dot{r}$ = v_o , RHS of equation (2) is called the impulsive of the force & the integral represents the impulse imported to the particle during the interval [t = t_0] and is equal to the change in the momentum of the particle. Again integrating through equation (3) once again w..r.to. Time, we get

$$m \dot{r} - m V_o = \int_{t_o}^{t} F dt^{\prime}$$

$$m \dot{r} = m V_o + \int_{t_o}^{t} F dt^{\prime}$$

$$\dot{r} = V_o + \int_{t_o}^{t} F dt^{\prime}$$

Now again Integrate on both sides

$$\int_{t_o}^{t} \dot{r} dt = \int_{t_o}^{t} V_o dt + \frac{1}{m} \int_{t_o}^{t} dt \int_{t_o}^{t} F dt^{\prime\prime}$$

$$[r]_{t_o}^{t} = V_o [t]_{t_o}^{t} + \frac{1}{m} \int_{t_o}^{t} dt \int_{t_o}^{t} F dt^{\prime\prime}$$

$$r - r_o = V_o [t - t_o] + \frac{1}{m} \int_{t_o}^{t} dt \int_{t_o}^{t} F dt^{\prime\prime} \quad \text{------- (4)}$$

Where r & r_o are the position of the particle at instants (T = t) & $(t = t_o)$ respectively.

If force is known and integration of equation (3) is carried,, we get explicite form of equation of the trajectory of the particle.

We shall consider some simple form of forces, these are

(i) F = ma ; a is constant

Example: *F = mg*

Where,
g – Acceleration due to gravity constant
These are constant forces

(ii) F = F (t) ; time dependent force
Alternating force of type F = $F_o Sin\omega t$, which is experience by a charged particle in alternating fields.

(iii) F = F (r) ; Position dependent force

Example: F = K. $\frac{r}{r^3}$

F = K. $\frac{1}{r^2}$

The force given by inverse square law, experienced in the gravitational and electrostatic fields.
(Or) F = - kr which experienced by particle performing linearly.
(iv) F = F ($\dot{r}$) ; Velocity dependent force
Example: F = KV or F = K V^n which represents frictional force experienced by particle moving in a viscous medium.

(v) Lorentz law of electromagnetic force
F = e [E + V x B]

1.8. MOTION UNDER A CONSTANT FORCE (OR) EQUATIONS OF MOTION FOR A CONSTANT FORCE

$\vec{F} = m\vec{a}$, a constant

For example: $\vec{F} = m\vec{g}$, when $\vec{g}$ be the acceleration due to earth's gravity can be considered constant. i.e., near the surface of the earths. So one of the familiar examples of motion under a constant force is motion under gravity.

When the applied force F on a particle is constant in time and hence there is a constant acceleration, then from Newton's second law of motion, acceleration is given by

$$\vec{a} = \frac{d\vec{V}}{dt}$$

$$\vec{a}dt = d\vec{v} \quad \text{---------------- (1)}$$

Integrate equation (1) on both sides under limits at time t = 0 to t and velocity V_0 to V, we get

$$\int_0^t \vec{a}\, dt = \int_{V_o}^{V} d\vec{V}$$

$$a(t)_0^t = (V)_{V_o}^{V}$$

$$a(t-0) = V - V_o$$

$$at = V - V_o$$

$$at + V_o = V$$

$$\boldsymbol{V = V_o + at} \quad \text{------------ (2)}$$

Second Equation (Distance): Consider V = $\frac{dr}{dt}$ or V = $\frac{dx}{dt}$

From equation (2)

$$V = V_o + at$$

$$\frac{dx}{dt} = V_o + at$$

$$dx = V_o dt + at\, dt \quad \text{------------ (3)}$$

Integrate equation (3) under limits at t = 0 to t and x_0 to x , we get

$$\int_{x_o}^{x} dx = \int_0^t V_o dt + \int_C^t at\, dt$$

$$\int_{x_o}^{x} dx = V_o \int_0^t dt + a \int_0^t t\, dt$$

$$(x)_{x_o}^{x} = V_o (t)_0^t + a \left(\frac{t^2}{2}\right)_0^t$$

$$(x - x_o) = V_o\,(t-0) + \frac{a}{2}\,(t^2 - 0)$$

$$x - x_o = V_o \mathrm{t} + \frac{1}{2}at^2$$

$$x = x_o + V_o \mathrm{t} + \frac{1}{2}at^2$$

When x_o = 0 Then

$$\left.\begin{array}{l} \boldsymbol{x = V_o t + \frac{1}{2}at^2} \\ \text{(OR)} \\ \boldsymbol{r = V_o t + \frac{1}{2}at^2} \\ \text{(OR)} \\ \boldsymbol{S = ut + \frac{1}{2}at^2} \end{array}\right\} \text{------------- (4)}$$

Third equation of motion:

The magnitude of acceleration is given by

$$\vec{a} = \frac{d\vec{V}}{dt}$$

Multiply and divide with dx

$$\vec{a} = \frac{d\vec{V}}{dt} X \frac{dx}{dx}$$

$$\vec{a} = \frac{dx}{dt} X \frac{d\vec{V}}{dx}$$

$$\vec{a} = V X \frac{d\vec{V}}{dx}$$

$$adx = Vdv \quad \text{----------------- (5)}$$

Integrate equation (5) under limits x_o to x and V_o to , we get

$$\int_{x_o}^{x} adx = \int_{V_o}^{V} VdV$$

$$a \int_{x_o}^{x} dx = \int_{V_o}^{V} VdV$$

$$a\,(x)_{x_o}^{x} = \left(\frac{V^2}{2}\right)_{V_o}^{V}$$

$$a\,(x - x_o) = \frac{1}{2}\left(V^2 - V_o^2\right)$$

When x_o = 0 , Then

$$ax = \frac{1}{2}\left(V^2 - V_o^2\right)$$

$$\left.\begin{array}{l} \boldsymbol{2\,ax = V^2 - V_o^2} \\ \text{(OR)} \\ \boldsymbol{2\,as = V_f^2 - V_i^2} \end{array}\right\} \text{----------- (6)}$$

Equations (2), (4) & (6) are the familiar equations that describe the translational motion of a particle in one dimension.

1.9. ATWOOD'S MACHINE

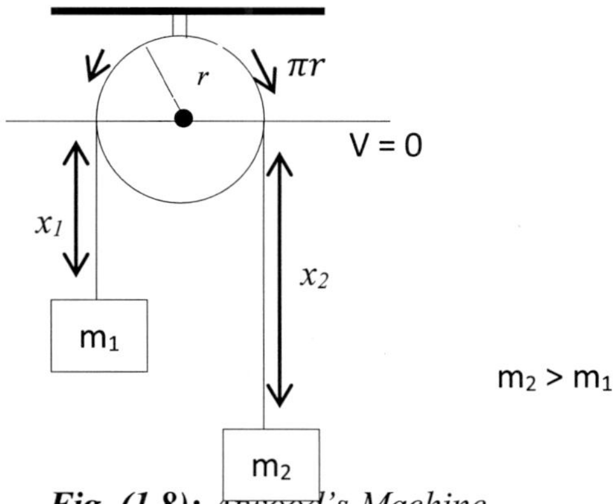

***Fig. (1.8):** Atwood's Machine*

Atwood's machine consists of 2 masses m_1 & m_2 suspended over a frictionless pulley of radius *'r'* and connected by an inextensible string of constant length *'l'* as shown in Fig (1.8). Since the masses can move only in vertical direction. This is one – dimensional problem.

As $m_2 > m_1$, velocity of $\frac{dx}{dt}$ of m_2 will be downwards, whereas velocity of m_1 will be upwards we can neglect the friction between the string and pulley. The acting on m_1 moving upwards.

From figure (1.8) it is clear that

$$l = x_1 + \pi r + x_2$$

$$x_2 = l - x_1 - \pi r \quad \text{-------------- (1)}$$

Differentiate equation (1) w.r.to. t

$$\dot{x}_2 = 0 - \dot{x}_1 - 0$$

$$\dot{x}_2 = -\dot{x}_1 \quad \text{-------------- (2)}$$

Kinetic Energy (K.E) of mass m_1 is given by

$$T_1 = \frac{1}{2} m_1 V_1^2$$

$$T_1 = \frac{1}{2} m_1 \dot{x}_1^2 \quad \text{---------- (3)} \left[V_1 = \frac{dx_1}{dt} = \dot{x}_1\right]$$

Kinetic Energy (K.E) of mass m_2 is given by

$$T_2 = \frac{1}{2} m_2 V_2^{\,2}$$

$$T_2 = \frac{1}{2} m_2 \dot{x}_2^{\,2} \qquad \text{---------- (4)} \quad \left[V_2 = \frac{dx_2}{dt} = \dot{x}_2\right]$$

Using Equation (2) in Equation (1), we get

$$T_2 = \frac{1}{2} m_2 (-\dot{x}_1)^2$$

$$T_2 = \frac{1}{2} m_2 \dot{x}_1^{\,2} \qquad \text{----------------- (5)}$$

Total kinetic Energy (K.E) of the system is given by

$$T = T_1 + T_2$$

$$T = \frac{1}{2} m_1 \dot{x}_1^{\,2} + \frac{1}{2} m_2 \dot{x}_1^{\,2}$$

$$T = \frac{1}{2} (m_1 + m_2)\, \dot{x}_1^{\,2} \qquad \text{--------------- (6)}$$

Potential Energy (P.E) of mass m_1 is given by

$$V_1 = -\, m_1 g x_1 \qquad \text{--------------- (7)} \qquad [P.E = -mgh]$$

Potential Energy (P.E) of mass m_1 is given by

$$V_2 = -\, m_2 g x_2 \qquad \text{--------------- (8)} \qquad [P.E = -mgh]$$

Using Equation (1) in Equation (8), we get

$$V_2 = -\, m_2 g\, (l - x_1 - \pi r) \qquad \text{-------------- (9)}$$

∴ Total P.E of the system is given by

$$V = V_1 + V_2$$

$$V = -\, m_1 g x_1 - m_2 g\, (l - x_1 - \pi r)$$

$$V = -\, m_1 g x_1 - m_2 g l + m_2 g\, x_1 + m_2 g \pi r$$

$$V = -\, m_1 g x_1 + m_2 g\, x_1 - m_2 g l + m_2 g \pi r$$

$$V = -\, g\, x_1 (m_1 - m_2) - m_2 g\, (l - \pi r)$$

$$V = -\, g\, x_1 (m_1 - m_2) + V_0 \qquad \text{-------------- (10)}$$

Where $V_0 = -\, m_2 g\, (l - \pi r)$ = Constant

Lagrangian L of the system is given by

$$L = T - V \qquad \text{-------------- (11)}$$

Using Equation (6) & (10) in Equation (11), we get

$$L = \frac{1}{2}(m_1 + m_2)\dot{x}_1^{\,2} - [-g\,x_1\,(m_1 - m_2) + V_0]$$

$$L = \frac{1}{2}(m_1 + m_2)\dot{x}_1^{\,2} + g\,x_1\,(m_1 - m_2) + V_0 \qquad \text{------- (12)}$$

Now Lagrange's Equation of motion is given by

$$\frac{d}{dt}\left(\frac{\partial L}{\partial \dot{x}_1}\right) - \frac{\partial L}{\partial x_1} = 0 \qquad \text{-------------------- (13)}$$

Partially differentiating Equation (12) w.r.to. $\dot{x}_1$ we get

$$\frac{\partial L}{\partial \dot{x}_1} = \frac{1}{2}(m_1 + m_2)\,2\,\dot{x}_1 + 0$$

$$\frac{\partial L}{\partial \dot{x}_1} = (m_1 + m_2)\,\dot{x}_1 \qquad \text{----------------- (14)}$$

Again partially differentiating Equation (12) w.r.to. x_1 we get

$$\frac{\partial L}{\partial x_1} = 0 + g\,(m_1 - m_2)$$

$$\frac{\partial L}{\partial x_1} = g\,(m_1 - m_2) \qquad \text{----------------- (15)}$$

Using Equation (14) & (15) in Equation (13), we get

$$\frac{d}{dt}(m_1 + m_2)\,\dot{x}_1 - g\,(m_1 - m_2) = 0$$

$$(m_1 + m_2)\,\ddot{x}_1 - g\,(m_1 - m_2) = 0$$

$$(m_1 + m_2)\,\ddot{x}_1 = g\,(m_1 - m_2)$$

$$\ddot{x}_1 = \frac{g\,(m_1 - m_2)}{m_1 + m_2} \qquad \text{------------------- (16)}$$

Cases:

(i) It should be noted that if $m_1 = m_2$, Then acceleration $\ddot{x}_1 = 0$ and two masses are stationary.

(ii) If one of the masses is very large then $\ddot{x}_1 \approx g$, which I negligible as compared to "$m_2 g$" and hence mass m_2 undergoes free-fall.

1.10. MOTION UNDER A TIME – DEPENDENT FORCE

$$\vec{F} = \vec{F}(t)$$

For example: Alternating force of the type $\vec{F} = \vec{F_o}Sin\omega t$, which is experienced by charged particles in alternating fields.
When the force acting on a particle is an explicit function of time, then equation of motion can be written as

$$m\frac{dV}{dt} = F(t)$$

$$dV = \frac{1}{m}F(t)dt$$

Assuming the initial conditions that V_0 to V at 0 to t, integration gives

$$\int_{V_o}^{V} dV = \frac{1}{m}\int_0^t F(t)dt$$

$$(V)_{V_o}^{V} = \frac{1}{m}\int_0^t F(t)dt$$

$$V - V_o = \frac{1}{m}\int_0^t F(t)dt$$

$$V = V_o + \frac{1}{m}\int_0^t F(t)dt$$

When $V_o = 0$, Then

$$\boldsymbol{V = \frac{1}{m}\int_0^t F(t)dt}$$

1.11. MOTION UNDER A VELOCITY DEPENDENT FORCE

It is very often the case that, in addition to constant applied forces, forces that depend on velocity are present. When a body is falling in a gravitational field, in additional to the gravitational force, there exists a force of resistance offered by air which is dependent on velocity. When bodies move through fluids, the viscous forces oppose the motion. For such systems, Newton's second law may be written in the form

$$m\frac{dV}{dt} = F(v) \qquad \text{------------ (1)}$$

$$m\frac{dV}{F(v)} = dt$$

Assuming the initial conditions that V_0 to V at t_0 to t, integration gives

$$m \int_{V_o}^{V} \frac{dV}{F(V)} = \int_{t_o}^{t} dt$$

$$m \int_{V_o}^{V} \frac{dV}{F(V)} = (t)_{t_o}^{t}$$

$$m \int_{V_o}^{V} \frac{dV}{F(V)} = t - t_o$$

$$\boldsymbol{t = t_o + m \int_{V_o}^{V} \frac{dV}{F(V)}} \qquad \text{-------------- (2)}$$

Multiply & Divide with dx to LHS of equation (1), we get

$$m \frac{dV}{dt} \times \frac{dx}{dx} = F(v)$$

$$m \frac{dV}{dx} \times \frac{dx}{dt} = F(v)$$

$$m \frac{dV}{dx} . V = F(v) \qquad \left[\begin{matrix} We\ know\ that \\ V = \frac{dx}{dt}\ Or\ V = \frac{dr}{dt} \end{matrix}\right]$$

$$m \frac{VdV}{F(V)} = dx \qquad \text{-------------- (3)}$$

Integrate on both sides with V_0 to V and x_0 to x, we get

$$m \int_{V_o}^{V} \frac{VdV}{F(v)} = \int_{x_o}^{x} dx$$

$$m \int_{V_o}^{V} \frac{VdV}{F(v)} = (x)_{x_o}^{x}$$

$$m \int_{V_o}^{V} \frac{VdV}{F(v)} = x - x_o$$

$$\left.\begin{matrix} \boldsymbol{x = x_o + m \int_{V_o}^{V} \frac{VdV}{F(v)}} \\ \text{(OR)} \\ \boldsymbol{r = r_o + m \int_{V_o}^{V} \frac{VdV}{F(v)}} \end{matrix}\right\} \qquad \text{-------------- (4)}$$

1.12. MOTION UNDER A POSITION DEPENDENT FORCE

$$\vec{F} = \vec{F}(\vec{r})$$

For Example:

- $\vec{F} = \frac{K\vec{r}}{r^3}$, the force given by inverse square law experienced in the gravitational or electrostatic field.
- $\vec{F} = \vec{r}$, which is experienced by a particle performing linear simple harmonic motion.

Forces that are functions of position occur frequently in nature. One example is the gravitational force. The equation of motion can be written as

$$m\,\frac{dV}{dt} = F(r)$$

$$m\,\frac{d}{dt}\left(\frac{dr}{dt}\right) = F(r) \quad \text{------ (1)} \quad \left[\begin{array}{c} We\ know\ that \\ V = \frac{dx}{dt}\ Or\ V = \frac{dr}{dt} \end{array}\right]$$

Multiply and Divide with dr to LHS of equation (1), we get

$$m\,\frac{d}{dt}\left(\frac{dr}{dt}\right) X\,\frac{dr}{dr} = F(r)$$

$$m\,\frac{d}{dt}\left(\frac{dr}{dt}\right).dr = F(r).dr \quad \text{----------- (2)}$$

Multiply and Divide with dt to LHS of equation (2), we get

$$m\,\frac{d}{dt}\left(\frac{dr}{dt}\right) X\,\frac{dr}{dt}.dt = F(r).dr \quad \text{------------ (3)}$$

Integrate equation (3) on both sides with r_0 to r and t_0 to t, we get

$$m\int_{t_o}^{t}\frac{d}{dt}\left(\frac{dr}{dt}\right).\frac{dr}{dt}dt = \int_{r_o}^{r}F(r).dr$$

$$m\int_{t_o}^{t}\frac{d}{dt}\left(\frac{dr}{dt}\right)^2 dt = \int_{r_o}^{r}F(r).dr$$

$$\frac{1}{2}m\left[\left(\frac{dr}{dt}\right)^2\right]_{t_o}^{t} = \int_{r_o}^{r}F(r).dr$$

$$\frac{1}{2}m\,[V^2]_{t_o}^{t} = \int_{r_o}^{r}F(r).dr \quad \text{------- (4)} \quad \left[\begin{array}{c} We\ know\ that \\ V = \frac{dx}{dt}\ Or\ V = \frac{dr}{dt} \end{array}\right]$$

Where r_o is the initial position at t_o

If V_0 is the initial velocity at t_o, then

$$\frac{1}{2} m \left(V^2 - V_o{}^2\right) = \int_{r_o}^{r} F(r).dr \qquad \text{------------ (5)}$$

If the force $\vec{F}$ is a conservative one, the work integral can always be evaluated, either explicitly or numerically.

$$\int_{r_o}^{r} F(r).dr = \int_{r_o}^{r} \nabla V(r).dr = V_{r_o} - V_r \qquad \text{---------- (6)}$$

Combining equation (5) & (6) the law of conservation of energy

$$\frac{1}{2} mV^2 - \frac{1}{2} mV_o{}^2 = V_{r_o} - V_r$$

$$\frac{1}{2} mV^2 + V_r = V_{r_o} + \frac{1}{2} mV_o{}^2 \qquad \text{-------------- (7)}$$

1.13. MOTION OF CHARGED PARTICLES IN MAGNETIC FIELDS

Consider a charged particle having a charge q, mass m and velocity V moving in a uniform magnetic field B. the force experienced by the charge is given by

$$F = qBVSin\theta$$

$$F = qBV_2 \qquad \text{--------------- (1)}$$

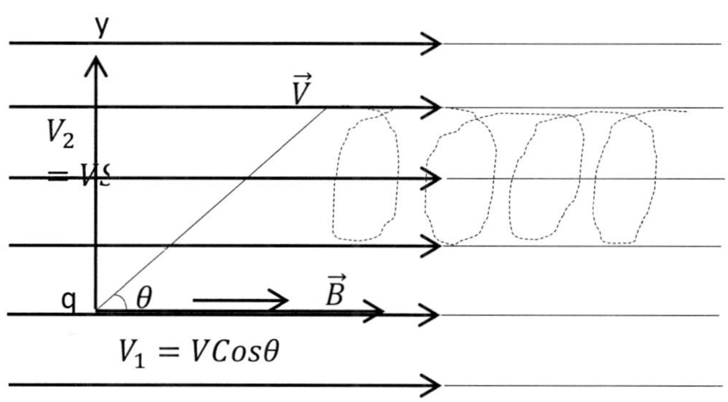

Fig. (1.9)

A necessary centripetal force act on charged particle

$$F = \frac{mV_2{}^2}{r} \qquad \text{-------------- (2)}$$

From equation (1) & (2), we have

$$qBV_2 = \frac{mV_2{}^2}{r}$$

$$r = \frac{mV_2{}^2}{qBV_2}$$

$$r = \frac{mV_2}{qB} \qquad \text{----------------- (3)}$$

From here $r \, \alpha \, V_2$

The time taken by the particle for one revolution. i.e. time taken to cover one complete circle is given by

$$Time = \frac{Distance}{Speed}$$

$$T = \frac{2\pi r}{V_2}$$

$$T = \frac{2\pi m V_2}{V_2 qB} \qquad \begin{bmatrix} From\ eq's(3) \\ r = \frac{mV_2}{qB} \end{bmatrix}$$

$$T = \frac{2\pi m}{qB} \qquad \text{-------------- (4)}$$

Thus the time taken is constant & independent of r because $r \, \alpha \, V_2$

From here, $T \, \alpha \, \frac{1}{B}$ i.e. if magnetic field is increases, time taken to complete one circle decreases

∴ $$Frequency \surd = \frac{1}{T}$$

$$\surd = \frac{qB}{2\pi m} \quad (From\ eq'n\ (4)) \qquad \text{------- (5)}$$

The radius r is sometimes called the gyro radius or Cyclotron radius. The angular velocity or angular frequency ω is given by

$$Angular\ Velocity\ \omega = 2\pi\surd$$

$$= \frac{2\pi qB}{2\pi m} \qquad (From\ eq'n\ (5))$$

$$\omega = \frac{qB}{m}$$

From above discussion we conclude the component $VSin\theta$ provide the circular motion whereas $VCos\theta$ provide straight line motion.

Due to this combined effect, motion of charge particle becomes helical.

Case (i):
If charge particle is thrown for to the magnetic field, then θ = 90^0

Here, $VSin\theta$ - Represents Circular Path
$VCos\theta$ - Represents straight path

$VSin\theta$	$VCos\theta$
V	0

So motion of particle becomes circular path

Case (ii):
If charge particle is thrown parallel to the magnetic field, then $\theta = 0^0$

$VSin\theta$	$VCos\theta$
0	V

So motion of particle becomes Straight path

CHAPTER – 2

CENTRAL FORCE MOTION

In this chapter we shall discuss the problem of two bodies moving under the influence of a mutual central force as an application of the Lagrangian formulation. Not all the problems of central force motion are integrable in terms of well – known functions. However, we shall attempt to explore the problem as thoroughly as is possible with the tools already developed.

2.1. INTRODUCTION

A central force is a force whose line of action is always directed towards a fixed point, called the centre or origin of the force, and whose magnitude depends only on the distance from the centre. If interaction between any two objects is represented by a central force, then the force is directed along the line joining the centres of the two objects. Central forces are important because we encounter them very often in physic. The familiar gravitational force is a central force. The electrostatic force between two charges is a central force. Even certain two-body nuclear interactions such as the scattering of α – particles by nuclei is governed by a central force. In this chapter, we shall discuss some of the salient features of central force motion.

CENTRAL FORCE MOTION

The motion of particles in force field can be mainly classified into two types

1) Bounded motion:

In this type of motion, the separating distance between two particles is fixed.

Examples:

i) The motion of planetary bodies around the sun.

ii) The motion of Electrons around the nucleus.

iii) The motion of Satellite around planets.

2) Unbounded motion:

In this type of motion, the separating distance between two particles is not fixed, both at the initial and final stages of the motion.

Example:

i) Scattering of alpha – particles

ii) Comets

Examples of Central Forces

Some real – life examples of central forces as follows;

i) Gravitational force of attraction, force acting between two masses.

ii) Electrostatic force of attraction or repulsion, force acting between two charges.

iii) Spring force, restoring force.

Characteristics of Central Force:

i) Central forces are long range forces, depending on the distance of separation, its magnitude decreases as the separation increases.

ii) It acts along a line joining the centre of two bodies that's why these are called central forces.

iii) Central forces are conservative in nature. This means work done by these forces is independent of the path traversed and depend only the initial and final points.

2.2. GENERAL PROPERTIES OF CENTRAL FORCE MOTION

The equation of motion for a particle of mass μ in central force field is given by

$$\mu\,\ddot{r} = F(r).\hat{r}$$

Where, $(r) = -\frac{\partial V}{\partial r}$, Since it is a vector equation, in effect we have 3 equations. We can learn a lot about the motion of the particle without actually solving these equations.

Angular Momentum

Let L be the angular momentum corresponding to the relative motion.

$$\therefore \qquad L = r\,X\,P \qquad \text{---------------} \ (1)$$

Differentiate equation (1) with respect to 't', we get

$$\frac{dL}{dt} = \frac{d}{dt}\,(r\,X\,P)$$

$$\frac{dL}{dt} = \left(\frac{dr}{dt}\,X\,P\right) + \left(r\,X\,\frac{dP}{dt}\right)$$

$$\frac{dL}{dt} = (V\,X\,P) + (r\,X\,F)$$

But V and P = μV are Parallel. So that $V \; X \; P = 0$

And for the central force case

$$r \; X \; F = F(r) X \, \hat{r} = 0$$

So that equation (2) becomes

$$\frac{dL}{dt} = 0 + 0$$

$$\frac{dL}{dt} = 0$$

L = Constant

Thus, in case of central force motion, the angular momentum is conserved, both in direction and magnitude.

Conservation of angular momentum implies that the relative motion occurs in a plane.

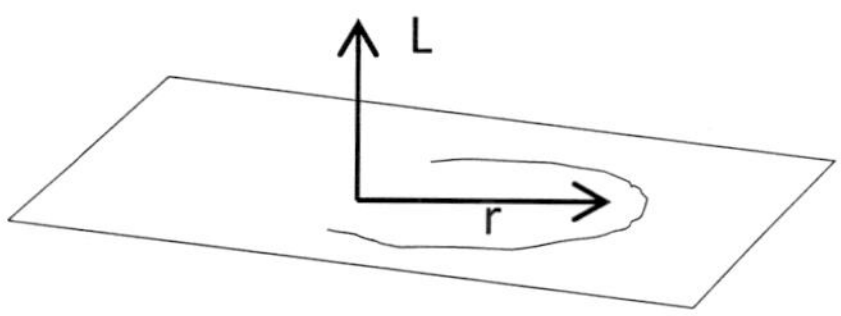

Fig. (2.2)

Here Direction of L is fixed and because r is perpendicular to L, So r must be in the same plane.

So, we can use plane polar co – ordinates (r, θ) to describe the motion.

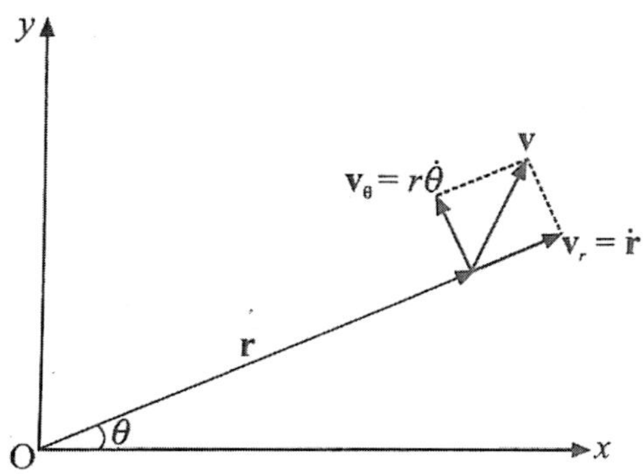

Fig. (2.3): *Motion of a particle in polar co – ordinates*

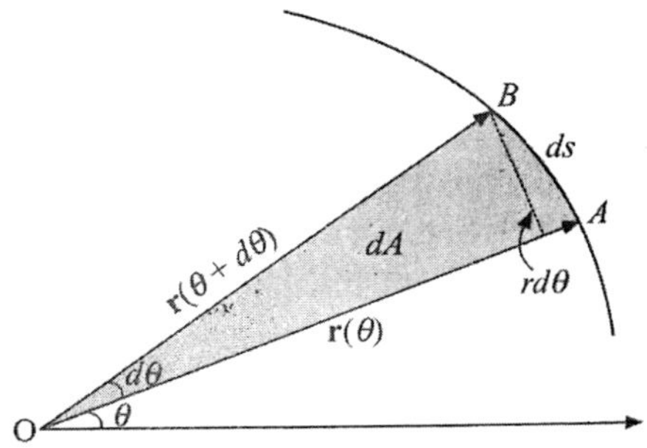

Fig. (2.4): *Area swept by a radius vector r in time dt*

We know that in plane polar co – ordinates

$$a = \ddot{r} = (\ddot{r} - r\,\dot{\theta}^2)\hat{r} + (2\dot{r}\dot{\theta} + r\ddot{\theta})\hat{\theta}$$

Therefore equation of motion $\ddot{r} = F(r)\hat{r}$, become

$$\mu\,(\ddot{r} - r\,\dot{\theta}^2)\hat{r} + \mu\,(2\dot{r}\dot{\theta} + r\ddot{\theta})\hat{\theta} = F(r)\hat{r}$$

On comparing both sides, we obtain following equations

$$\mu\,(\ddot{r} - r\,\dot{\theta}^2) = F(r)$$

$$\mu\,(2\dot{r}\dot{\theta} + r\ddot{\theta}) = 0$$

By multiplying second equation on both sides by r, we obtain

$$\frac{d}{dt}(\mu r^2\dot{\theta}) = 0$$

This equation yields

$$\mu r^2\dot{\theta} = \text{L (Constant)}$$

We called this L is constant because it is nothing but the angular momentum of the particle about the origin.

Note that $L = l\omega$, with $l = \mu r^2$

Law of Equal Areas

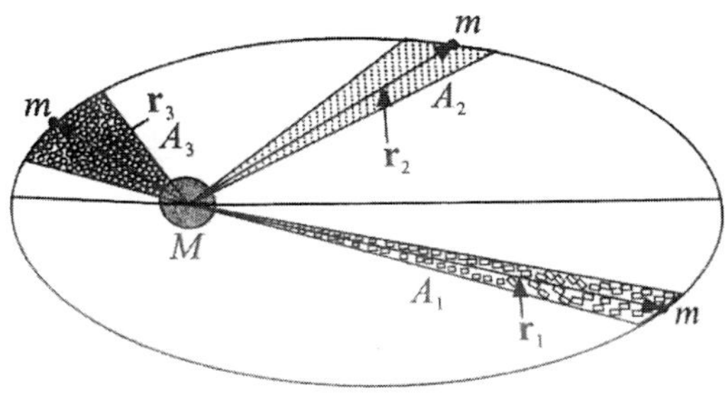

Fig. (2.5): *Law of Equal Areas:* $A_1 = A_2 = A_3$

As the particle moves along the trajectory so that the angle θ changes by an infinitesimal amount $d\theta$, the area swept with respect to the origin is

$$dA = \frac{1}{2}r^2 d\theta$$

$$\frac{dA}{dt} = \frac{1}{2}r^2\dot{\theta} \quad \text{---------------} \ (3)$$

Which is the rate at which the radius vector sweeps out the area. Substituting the value of $\dot{\theta} = \frac{L}{\mu r^2}$ in equation (3)

$$\frac{dA}{dt} = \frac{L}{2\mu} = \text{Constant}$$

Because here L is Constant

Thus constancy of areal velocity is a property of all central forces, not just the gravitational forces. And it holds due to conservation of angular momentum.
Kinetic Energy in plane polar co –ordinates can be written as.

$$K = \frac{1}{2}\ \mu V.V$$

$$K = \frac{1}{2}\ \mu\ (\dot{r}\hat{r} + r\dot{\theta}\hat{\theta}).\ (\dot{r}\hat{r} + r\dot{\theta}\hat{\theta})$$

$$K = \frac{1}{2}\ \mu\ \dot{r}^2 + \frac{1}{2}\ \mu\ r^2\ \dot{\theta}^2$$

Potential energy $V(r)$ can be obtained by the basic formula.

$$V(r) - V(r_o) = -\ \int_0^r F(r)\ dr$$

Where r_o denotes the location of a reference point.

Total energy E from work – energy theorem

$$E = \frac{1}{2}\,\mu\,\dot{r}^2 + \frac{1}{2}\,\mu\,r^2\dot{\theta}^2 + V(r) = Constant$$

We have

$$L = \mu\,r^2\dot{\theta}$$

$$\frac{1}{2}\,\mu\,r^2\,\dot{\theta}^2 = \frac{L^2}{2\mu r^2}$$

So that

$$E = \frac{1}{2}\,\mu\,\dot{r}^2 + \frac{L^2}{2\mu r^2} + V(r) \qquad \text{-------------- (4)}$$

Note that the expression for E does not contain $\dot{\theta}$

2.3. REDUCTION OF TWO – BODY PROBLEM TO EQUIVALENT ONE – BODY PROBLEM

Let us consider a system of two particles of masses m_1 & m_2 having position vectors $\vec{r}_1$ & $\vec{r}_2$ with respect to origin 'O' of an inertial frame of reference. Let $\vec{r}$ be the position vector of m_1 w.r.to. m_2.

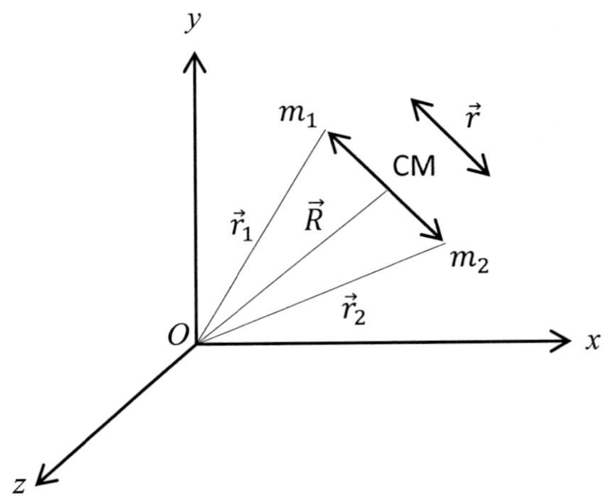

Fig. (2.1)

From Fig. (2.1) it is clear that

$$\left.\begin{aligned}\vec{r}_2 &= \vec{r} + \vec{r}_1 \\ \vec{r}_1 &= \vec{r}_2 - \vec{r}\end{aligned}\right\} \qquad \text{-------------- (1)}$$

The Lagrangian (L) of the system is given by

$$L = T - V$$

Where,

T – is kinetic energy (K.E)

V – is potential energy of system which is function of scalar distance *(r)* between two particles.

$$L = \frac{1}{2}\, m_1\, \dot{\vec{r}}_1^{\,2} + \frac{1}{2}\, m_2\, \dot{\vec{r}}_2^{\,2} - U_{(r)} \quad \text{------------ (2)}$$

Let $\vec{R}$ be the position vector of center of mass of two particles then by definition

$$\vec{R} = \frac{m_1\, \vec{r}_1 + m_2\, \vec{r}_2}{m_1 + m_2} \quad \text{-------------------- (3)}$$

From equation (1)

$$\vec{r}_2 = \vec{r} + \vec{r}_1 \quad \text{------------------------ (4)}$$

Using equation (4) in equation (3), we get

$$\vec{R} = \frac{m_1\, \vec{r}_1 + m_2\, (\vec{r} + \vec{r}_1)}{m_1 + m_2}$$

$$\vec{R}\,(m_1 + m_2) = m_1\, \vec{r}_1 + m_2\, (\vec{r} + \vec{r}_1)$$

$$\vec{R}\,(m_1 + m_2) = m_1\, \vec{r}_1 + m_2\, \vec{r} + m_1\, \vec{r}_1$$

$$\vec{R}\,(m_1 + m_2) = (m_1 + m_2)\vec{r}_1 + m_2\, \vec{r}$$

$$(m_1 + m_2)\vec{r}_1 = \vec{R}\,(m_1 + m_2) - m_2\, \vec{r}$$

$$\vec{r}_1 = \frac{\vec{R}\,(m_1 + m_2)}{(m_1 + m_2)} - \frac{m_2\, \vec{r}}{(m_1 + m_2)}$$

$$\vec{r}_1 = \vec{R} - \frac{m_2\, \vec{r}}{(m_1 + m_2)}$$

$$\dot{\vec{r}}_1 = \dot{\vec{R}} - \frac{m_2\, \dot{\vec{r}}}{(m_1 + m_2)} \quad \text{--------------- (5)}$$

From equation (1)

$$\vec{r}_1 = \vec{r}_2 - \vec{r} \quad \text{-------------- (6)}$$

Using equation (6) in equation (3), we get

$$\vec{R} = \frac{m_1\, (\vec{r}_2 - \vec{r}) + m_2\, \vec{r}_2}{m_1 + m_2}$$

$$\vec{R}\,(m_1 + m_2) = m_1\,(\vec{r}_2 - \vec{r}) + m_2\,\vec{r}_2$$

$$\vec{R}\,(m_1 + m_2) = m_1\,\vec{r}_2 - m_1\,\vec{r} + m_2\,\vec{r}_2$$

$$\vec{R}\,(m_1 + m_2) + m_1\,\vec{r} = (m_1 + m_2)\vec{r}_2$$

$$\vec{r}_2 = \frac{\vec{R}\,(m_1 + m_2)}{(m_1 + m_2)} + \frac{m_1\,\vec{r}}{(m_1 + m_2)}$$

$$\vec{r}_2 = \vec{R} + \frac{m_1\,\vec{r}}{(m_1 + m_2)}$$

$$\dot{\vec{r}}_2 = \dot{\vec{R}} + \frac{m_1\,\dot{\vec{r}}}{(m_1 + m_2)} \qquad \text{--------------- (7)}$$

Put equation (5) & equation (7) Values in equation (2), we get

$$L = \frac{1}{2}\,m_1\left[\dot{\vec{R}} - \frac{m_2\,\dot{\vec{r}}}{(m_1 + m_2)}\right]^2 + \frac{1}{2}\,m_2\left[\dot{\vec{R}} + \frac{m_1\,\dot{\vec{r}}}{(m_1 + m_2)}\right]^2 - V_{(r)}$$

$$L = \frac{1}{2}\,m_1\left[\dot{\vec{R}}^2 + \frac{{m_2}^2\,\dot{\vec{r}}^2}{(m_1 + m_2)^2} - \frac{2\,\dot{\vec{R}}\,m_2\,\dot{\vec{r}}}{(m_1 + m_2)}\right] + \frac{1}{2}\,m_2\left[\dot{\vec{R}}^2 + \frac{{m_1}^2\,\dot{\vec{r}}^2}{(m_1 + m_2)^2} + \frac{2\,\dot{\vec{R}}\,m_1\,\dot{\vec{r}}}{(m_1 + m_2)}\right] - V_{(r)}$$

$$L = \frac{1}{2}\,m_1\,\dot{\vec{R}}^2 + \frac{m_1\,{m_2}^2\,\dot{\vec{r}}^2}{2(m_1 + m_2)^2} - \frac{m_1\,\dot{\vec{R}}\,m_2\,\dot{\vec{r}}}{(m_1 + m_2)} + \frac{1}{2}\,m_2\,\dot{\vec{R}}^2 + \frac{m_2\,{m_1}^2\,\dot{\vec{r}}^2}{2(m_1 + m_2)^2} + \frac{2\,\dot{\vec{R}}\,m_2\,\dot{\vec{r}}}{(m_1 + m_2)} - V_{(r)}$$

$$L = \frac{1}{2}\,m_1\,\dot{\vec{R}}^2 + \frac{m_1\,{m_2}^2\,\dot{\vec{r}}^2}{2(m_1 + m_2)^2} + \frac{1}{2}\,m_2\,\dot{\vec{R}}^2 + \frac{m_2\,{m_1}^2\,\dot{\vec{r}}^2}{2(m_1 + m_2)^2} - V_{(r)}$$

$$L = \frac{1}{2}\,m_1\,\dot{\vec{R}}^2 + \frac{1}{2}\,m_2\,\dot{\vec{R}}^2 + \frac{m_1\,{m_2}^2\,\dot{\vec{r}}^2}{2(m_1 + m_2)^2} + \frac{m_2\,{m_1}^2\,\dot{\vec{r}}^2}{2(m_1 + m_2)^2} - V_{(r)}$$

$$L = \frac{1}{2}\,(m_1 + m_2)\,\dot{\vec{R}}^2 + \frac{m_1\,m_2\,(m_2 + m_1)\,\dot{\vec{r}}^2}{2(m_1 + m_2)^2} - V_{(r)}$$

$$L = \frac{1}{2}\,(m_1 + m_2)\,\dot{\vec{R}}^2 + \frac{1}{2}\left(\frac{m_1\,m_2}{m_1 + m_2}\right)\dot{\vec{r}}^2 - V_{(r)}$$

$$L = \frac{1}{2}\,M\,\dot{\vec{R}}^2 + \frac{1}{2}\,\mu\,\dot{\vec{r}}^2 - V_{(r)} \qquad \text{----------------- (8)}$$

Where,

$M = m_1 + m_2$ = Total mass of system

$\mu = \frac{m_1 m_2}{m_1 + m_2}$ = Reduced mass of the system

As center of mass is either at rest or moving uniformly,
$\vec{\dot{R}}$ = Constant = 0

Hence equation (8) reduces to

$$\boldsymbol{L = \frac{1}{2}\, \mu \vec{\dot{r}}^2 - V_{(r)}} \quad \text{-------------- (9)}$$

This Lagrangian is exactly the same as that of a particle having mass μ, moving at a distance $\vec{r}$ w.r.to. Origin 'O'.

The equation of motion for the system is given by

$$\frac{d}{dt}\left(\frac{\partial L}{\partial \dot{r}}\right) - \frac{\partial L}{\partial r} = 0 \quad \text{-------------- (10)}$$

From equation (9)

$$\frac{\partial L}{\partial \dot{r}} = \frac{\partial}{\partial \dot{r}}\left(\frac{1}{2}\,\mu \vec{\dot{r}}^2 - V_{(r)}\right) \quad \& \quad \frac{\partial L}{\partial r} = \frac{\partial}{\partial r}\left(\frac{1}{2}\,\mu \vec{\dot{r}}^2 - V_{(r)}\right)$$

$$\frac{\partial L}{\partial \dot{r}} = \frac{1}{2}\, 2\, \mu\, \dot{r} \qquad \& \quad \frac{\partial L}{\partial r} = -\frac{\partial V}{\partial r} \quad \text{------ (11b)}$$

$$\frac{\partial L}{\partial \dot{r}} = \mu\, \dot{r} \quad \text{------------ (11a)}$$

Put the values of $\frac{\partial L}{\partial \dot{r}}$ & $\frac{\partial L}{\partial r}$ in equation (10), we get

$$\frac{d}{dt}\,(\mu\, \dot{r}) - \left(-\frac{\partial V}{\partial r}\right) = 0$$

$$\frac{d}{dt}\,(\mu\, \dot{r}) + \frac{\partial V}{\partial r} = 0$$

$$\mu\, \ddot{r} + \frac{\partial V}{\partial r} = 0$$

$$\mu\, \ddot{r} = -\frac{\partial V}{\partial r} = f(r)$$

Where,

$$f(r) = -\frac{\partial V}{\partial r} = \text{Conservative force}$$

2.4. EFFECTIVE POTENTIAL

Equation (4) is identified to the total energy expression of a particle of mass μ moving under the influence of an effective potential

$$V_{eff} = V(r) + \frac{L^2}{2\mu r^2} \qquad \text{------------------- (5)}$$

The effective potential is the sum of the potential V(r) and an additional term called the centrifugal potential defined as

$$V_C = \frac{L^2}{2\mu r^2} \qquad \text{------------------------- (6)}$$

This V_{eff} is equivalent to a force term

$$F_{eff}(r) = F(r) + \frac{L^2}{\mu r^3} \qquad \text{-------------- (7)}$$

The additional force term $\frac{L^2}{\mu r^3}$, a fictitious force, is usually referred to as the centrifugal force, F_C :

$$F_C = \frac{L^2}{\mu r^3} = \mu r\dot{\theta}^2 \qquad \text{--------------------- (8)}$$

Since L is constant, it is evident that F_C is a also a central force, with this effective force the equation of motion is identical to that of a one – dimensional one:

$$\mu\ddot{r} = F_{eff} \qquad \text{------------------------ (9)}$$

In physical terms equation (9) describes the motion of a particle of mass μ under a central force as viewed by an observer at the centre of force from s rotating reference frame. In this rotating frame the force seems to be $F_{eff}(r)$.

2.5. THE EQUATION OF MOTION AND FIRST INTEGRALS

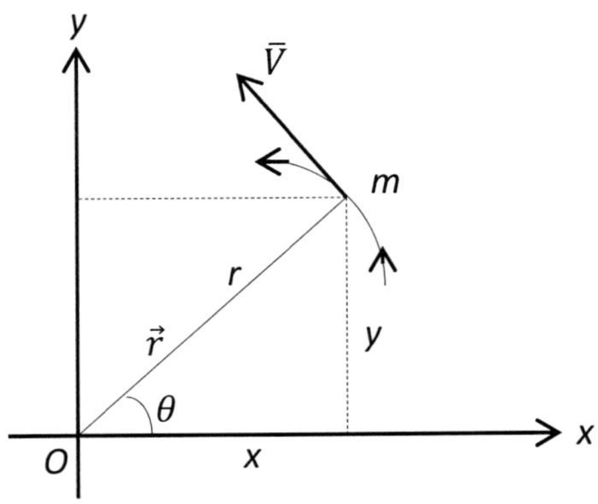

Fig. (2.6)

Consider a particle of mass 'm' moving under central force field $\vec{F}$ is given by

$$\vec{F} = \vec{n}\,F(r)$$

$$\vec{F} = \vec{n}\left(-\frac{dV}{dr}\right) \quad \text{--------------------- (1)}$$

Where,

$\vec{n} = \frac{\vec{r}}{r}$ = unit vector along $\vec{r}$

$\vec{r}$ = Position vector of particle w.r.to. Origin 'O'

The Lagrangian (L) of the system is given by

L = T – V ------------------------ (2)

Where,

T = Kinetic Energy (K.E) of Particle
V = Potential Energy (P.E) of particle.

Now, $$T = \frac{1}{2}mV_x{}^2 + \frac{1}{2}mV_y{}^2$$

$$T = \frac{1}{2}m\dot{x}^2 + \frac{1}{2}m\dot{y}^2 \quad \text{------- (3)} \qquad \left(\begin{matrix} We\ know\ that \\ V_x = \frac{dx}{dt} = \dot{x} \\ V_y = \frac{dy}{dt} = \dot{y} \end{matrix}\right)$$

And V = V(r) ---------------- (4)

Put the values of T & V in equation (2), we get

$$L = \frac{1}{2}m\dot{x}^2 - \frac{1}{2}m\dot{y}^2 - V(r)$$

$$L = \frac{1}{2}m\,(\dot{x}^2 + \dot{y}^2) - V(r) \quad \text{---------------- (5)}$$

Now we have to convert $\dot{x}^2$ & $\dot{y}^2$ in the form of polar coordinates (r, θ)

From Fig. (2.6), it is clear that

$$Cos\theta = \frac{Base}{Hypotenuse} = \frac{x}{r}$$

$$x = r\ cos\theta$$

Differentiate on both sides w. r. to. Time, we get

$$\dot{x} = r\,(-sin\theta).\dot{\theta} + cos\theta.\dot{r}$$

$$\dot{x} = -rsin\theta.\dot{\theta} + cos\theta.\dot{r} \quad \text{---------------- (6)}$$

Squaring on both sides of equation (6)

$$(\dot{x})^2 = \left(-rsin\theta.\dot{\theta} + cos\theta.\dot{r}\right)^2$$

$$\dot{x}^2 = r^2sin^2\theta.\dot{\theta}^2 + cos^2\theta.\dot{r}^2 - 2rsin\theta.\dot{\theta}\,cos\theta.\dot{r} \quad \text{--------- (7)}$$

From Fig (2.6), it is clear that

$$Sin\theta = \frac{Perpendicular}{Hypotenuse}$$

$$Sin\theta = \frac{y}{r}$$

$$y = r\,sin\theta$$

Differentiate on both sides w. r. to. Time, we get

$$\dot{y} = rcos\theta.\dot{\theta} + sin\theta.\dot{r}$$

Squaring on both sides, we get

$$(\dot{y})^2 = \left(rcos\theta.\dot{\theta} + sin\theta.\dot{r}\right)^2$$

$$\dot{y}^2 = r^2cos^2\theta.\dot{\theta}^2 + sin^2\theta.\dot{r}^2 + 2rcos\theta.\dot{\theta}\,sin\theta.\dot{r} \quad \text{--------- (8)}$$

Now add equation (7) and (8), we get

$$\dot{x}^2 + \dot{y}^2 = r^2sin^2\theta.\dot{\theta}^2 + cos^2\theta.\dot{r}^2 - 2rsin\theta.\dot{\theta}\,cos\theta.\dot{r} + r^2cos^2\theta.\dot{\theta}^2 + sin^2\theta.\dot{r}^2 + 2rcos\theta.\dot{\theta}\,sin\theta.\dot{r}$$

$$\dot{x}^2 + \dot{y}^2 = r^2sin^2\theta.\dot{\theta}^2 + cos^2\theta.\dot{r}^2 + r^2cos^2\theta.\dot{\theta}^2 + sin^2\theta.\dot{r}^2$$

$$\dot{x}^2 + \dot{y}^2 = r^2sin^2\theta.\dot{\theta}^2 + r^2cos^2\theta.\dot{\theta}^2 + cos^2\theta.\dot{r}^2 + sin^2\theta.\dot{r}^2$$

$$\dot{x}^2 + \dot{y}^2 = r^2\dot{\theta}^2\,(sin^2\theta + cos^2\theta) + \dot{r}^2\,(cos^2\theta + sin^2\theta)$$

$$\dot{x}^2 + \dot{y}^2 = r^2\dot{\theta}^2\,(1) + \dot{r}^2\,(1)$$

$$\dot{x}^2 + \dot{y}^2 = r^2\dot{\theta}^2 + \dot{r}^2 \quad \text{------------- (9)}$$

Put the value of $\dot{x}^2 + \dot{y}^2$ in equation (5), we get

$$L = \frac{1}{2}m\,(\dot{x}^2 + \dot{y}^2) - V(r) \quad \text{------------- (5)}$$

$$L = \frac{1}{2}m\,(r^2\dot{\theta}^2 + \dot{r}^2) - V(r) \qquad \text{------------- (10)}$$

$$L = \frac{1}{2}m\,r^2\dot{\theta}^2 + \frac{1}{2}m\,\dot{r}^2 - V(r) \qquad \text{------------- (11)}$$

Now Lagrangian equation of motion for the system is given by

$$\frac{d}{dt}\left(\frac{\partial L}{\partial \dot{q}_k}\right) - \frac{\partial L}{\partial q_k} = 0 \qquad \text{---------------- (12)}$$

This is the basic form of Lagrangian equation.
Here q_k is the generalized co-ordinate

But in equation (11) we have two generalized co-ordinates, $r\ and\ \theta$. If we know $r\ and\ \theta$ then we can easily locate the position of the particle in the system.

Now Lagrangian equation of motion for the system in terms of r is given by (Replace q_k by r in equation 12)

$$\therefore \qquad \frac{d}{dt}\left(\frac{\partial L}{\partial \dot{r}}\right) - \frac{\partial L}{\partial r} = 0 \qquad \text{---------------- (13)}$$

From equation (11), we get

$$\frac{\partial L}{\partial \dot{r}} = \frac{\partial}{\partial \dot{r}}\left(\frac{1}{2}m\,r^2\dot{\theta}^2 + \frac{1}{2}m\,\dot{r}^2 - V(r)\right)$$

$$\frac{\partial L}{\partial \dot{r}} = 0 + \frac{1}{2}m\,2\,\dot{r}$$

$$\frac{\partial L}{\partial \dot{r}} = m\dot{r} \qquad \text{------------------ (a)}$$

And $$\frac{\partial L}{\partial r} = \frac{\partial}{\partial r}\left(\frac{1}{2}m\,r^2\dot{\theta}^2 + \frac{1}{2}m\,\dot{r}^2 - V(r)\right)$$

$$\frac{\partial L}{\partial r} = \frac{1}{2}m\,2\,r\,\dot{\theta}^2 + 0 - \frac{\partial V}{\partial r}$$

$$\frac{\partial L}{\partial r} = m\,r\,\dot{\theta}^2 - \frac{\partial V}{\partial r} \qquad \text{------------------ (b)}$$

Put the values of equation (a) & (b) in equation (13), we get

$$\frac{d}{dt}(m\dot{r}) - \left[mr\dot{\theta}^2 - \frac{\partial V}{\partial r}\right] = 0$$

$$m\ddot{r} - mr\dot{\theta}^2 - \left(-\frac{\partial V}{\partial r}\right) = 0$$

$$m\ddot{r} - mr\dot{\theta}^2 - F(r) = 0 \quad \text{---------} (14) \quad \left[F(r) = -\frac{\partial V}{\partial r}\right]$$

Now Lagrangian equation of motion for the system in terms of θ, is given by (Replace q_k by θ in equation 12)

$$\therefore \quad \frac{d}{dt}\left(\frac{\partial L}{\partial \dot{\theta}}\right) - \frac{\partial L}{\partial \theta} = 0 \quad \text{----------------} (15)$$

From equation (11), we get

$$\frac{\partial L}{\partial \dot{\theta}} = \frac{\partial}{\partial \dot{\theta}}\left(\frac{1}{2}m\,r^2\dot{\theta}^2 + \frac{1}{2}m\,\dot{r}^2 - V(r)\right)$$

$$\frac{\partial L}{\partial \dot{\theta}} = \frac{1}{2}mr^2 2\,\dot{\theta}$$

$$\frac{\partial L}{\partial \dot{\theta}} = mr^2\,\dot{\theta} \quad \text{---------------} (c)$$

And

$$\frac{\partial L}{\partial \theta} = \frac{\partial}{\partial \theta}\left(\frac{1}{2}m\,r^2\dot{\theta}^2 + \frac{1}{2}m\,\dot{r}^2 - V(r)\right)$$

$$\frac{\partial L}{\partial \theta} = 0 \quad \text{---------------} (d)$$

Put the values of equation (c) & (d) in equation (15), we get

$$\frac{d}{dt}\left(mr^2\dot{\theta}\right) - 0 = 0$$

$$\frac{d}{dt}\left(mr^2\dot{\theta}\right) = 0 \quad \text{-------------------} (16)$$

$$\frac{d}{dt}(I\,\omega) = 0 \quad \left[\begin{array}{c} We\ know\ that \\ I = mr^2\ (Inertia) \\ \dot{\theta} = \frac{d\theta}{dt} = \omega\ (Angular\ Momentum \end{array}\right]$$

$$\frac{d}{dt}\left(\vec{L}\right) = 0$$

$$\vec{L} = Constant = mr^2\dot{\theta}^2 \quad \text{-----------------} (17)$$

Equation (17) is known as first integral of motion.

Areal Velocity

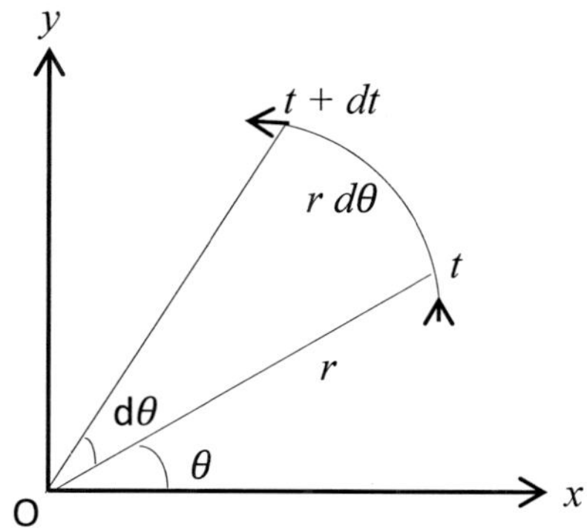

Fig. (2.7)

Area swept out by radius vector per unit time, the small area (dA) swept by radius vector *r* in small time *dt* is given by

$$dA = \frac{1}{2}\ X\ Base\ X\ Height$$

$$dA = \frac{1}{2}\ X\ r\ X\ rd\theta$$

$$dA = \frac{1}{2}\ r^2 d\theta$$

Areal Velocity $\quad \frac{dA}{dt} = \frac{1}{2}\ r^2\ \frac{d\theta}{dt}$

$$\frac{dA}{dt} = \frac{1}{2}\ r^2\ \dot{\theta} \qquad \text{----------------- (18)}$$

From equation (16)

$$\frac{d}{dt}\left(mr^2\dot{\theta}\right) = 0$$

Here m is constant

$$m\frac{d}{dt}\left(r^2\dot{\theta}\right) = 0$$

$$\frac{d}{dt}\left(r^2\dot{\theta}\right) = \frac{0}{m}$$

$$\frac{d}{dt}\left(r^2\dot{\theta}\right) = 0$$

Multiply $\frac{1}{2}$ on both sides, we get

$$\frac{d}{dt}\left(\frac{1}{2}r^2\dot{\theta}\right) = \frac{1}{2}\ X\ 0$$

$$\frac{d}{dt}\left(\frac{1}{2}r^2\dot{\theta}\right) = 0 \qquad \text{--------------- (19)}$$

$\therefore$ $$\frac{1}{2}r^2\dot{\theta} = \text{Constant} \qquad \text{--------------- (20)}$$

$\therefore$ Areal velocity = Constant

The conservation of $\vec{L}$ is thus equivalent to saying that areal velocity is constant. Equation (20) is known as Kepler's 2nd law of planetary motion

As $mr^2\dot{\theta}$ = L

Squaring on both sides, we get

$$m^2\, r^4\, \dot{\theta}^2 = L^2$$

$$\left(mr\dot{\theta}^2\right)(mr^3) = L^2$$

$$mr\dot{\theta}^2 = \frac{L^2}{mr^3} \qquad \text{--------------- (21)}$$

As $m\ddot{r} - mr\dot{\theta}^2 - F(r) = 0$

Put the value of equation (21) in above equation, we get

$$m\ddot{r} - \frac{L^2}{mr^3} - F(r) = 0$$

$$m\ddot{r} = \frac{L^2}{mr^3} + F(r)$$

$$m\ddot{r} = \frac{L^2}{mr^3} - \frac{\partial V}{\partial r} \qquad [F(r) = -\partial V/\partial r]$$

$$m\ddot{r} = -\frac{1}{2}\frac{d}{dr}\left(\frac{L^2}{mr^2}\right) - \frac{\partial V}{\partial r}$$

$$m\ddot{r} = -\frac{1}{2}\frac{d}{dr}\left(\frac{L^2}{mr^2}\right) - \frac{dV}{dr}$$

$$\left[-\frac{1}{2}\frac{d}{dr}\left(\frac{L^2}{mr^2}\right)\right]$$
$$= -\frac{1}{2}\frac{L^2}{m}\frac{d}{dr}(r^{-2})$$
$$= -\frac{1}{2}\frac{L^2}{m}\frac{d}{dr}(2r^{-3})$$
$$= \frac{L^2}{mr^2}$$

$$m\ddot{r} = -\frac{d}{dr}\left[\frac{L^2}{2mr^2} + V\right]$$

Multiply both sides by $\dot{r}$ or $\frac{dr}{dt}$

$$m\ddot{r}\dot{r} = -\frac{d}{dr}\left[\frac{L^2}{2mr^2} + V\right].\dot{r}$$

$$m\ddot{r}\dot{r} = -\frac{d}{dr}\left[\frac{L^2}{2mr^2} + V\right].\frac{dr}{dt}$$

$$m\ddot{r}\dot{r} = -\frac{d}{dt}\left[\frac{L^2}{2mr^2} + V\right]$$

$$\frac{1}{2}\frac{d}{dt}[m\dot{r}^2] = -\frac{d}{dt}\left[\frac{L^2}{2mr^2} + V\right] \qquad \left[\begin{array}{l}\frac{1}{2}\frac{d}{dt}m\dot{r}^2 \\ = \frac{1}{2}m2\dot{r} \\ = m\dot{r}\ddot{r}\end{array}\right]$$

$$\frac{d}{dt}\left[\frac{1}{2}m\dot{r}^2\right] + \frac{d}{dt}\left[\frac{L^2}{2mr^2} + V\right] = 0$$

$$\frac{d}{dt}\left[\frac{1}{2}m\dot{r}^2 + \frac{L^2}{2mr^2} + V\right] = 0$$

$$\frac{1}{2}m\dot{r}^2 + \frac{L^2}{2mr^2} + V = \text{Constant} \qquad \text{------------ (22)}$$

We know that $L = mr^2\dot{\theta}$

$$\frac{1}{2}m\dot{r}^2 + \frac{(mr^2\dot{\theta})^2}{2mr^2} + V = \text{Constant}$$

$$\frac{1}{2}m\dot{r}^2 + \frac{m^2r^4\dot{\theta}^2}{2mr^2} + V = \text{Constant}$$

$$\frac{1}{2}m\dot{r}^2 + \frac{1}{2}mr^2\dot{\theta}^2 + V = \text{Constant}$$

$$\frac{1}{2}m\,(\dot{r}^2 + r^2\dot{\theta}^2) + V = \text{Constant}$$

$$\frac{1}{2}m\,(\dot{r}^2 + r^2\dot{\theta}^2) + V = \text{E} = \text{Constant} \qquad \text{------- (23)}$$

This is the statement of conservation of energy.

From equation (22)

$$\frac{1}{2}m\dot{r}^2 + \frac{L^2}{2mr^2} + V = \text{E}$$

$$\frac{1}{2}m\dot{r}^2 = \text{E} - \frac{L^2}{2mr^2} - V$$

$$\dot{r}^2 = \frac{2}{m}\left[\mathrm{E} - \frac{L^2}{2mr^2} - V\right]$$

$$\dot{r} = \left[\frac{2}{m}\left(E - V - \frac{L^2}{2mr^2}\right)\right]^{1/2}$$

$$\frac{dr}{dt} = \left[\frac{2}{m}\left(E - V - \frac{L^2}{2mr^2}\right)\right]^{1/2}$$

$$dr = \frac{dr}{\left[\frac{2}{m}\left(E - V - \frac{L^2}{2mr^2}\right)\right]^{1/2}}$$

Now integrating on both sides over limits r_o to r

$$\int dt = \int_{r_0}^{r} \frac{dr}{\left[\frac{2}{m}\left(E - V - \frac{L^2}{2mr^2}\right)\right]^{1/2}}$$

$$t = \int_{r_0}^{r} \frac{dr}{\left[\frac{2}{m}\left(E - V - \frac{L^2}{2mr^2}\right)\right]^{1/2}} \qquad \text{---------------- (24)}$$

Above equation gives 't' as a function of 'r' and constant E, L, r_0

As $\qquad L = mr^2\dot{\theta}$

$$\dot{\theta} = \frac{L}{mr^2}$$

$$\frac{d\theta}{dt} = \frac{L}{mr^2}$$

$$d\theta = \frac{Ldt}{m\,r^2}$$

Integrating on both sides, we get

$$\int_{\theta_0}^{\theta} d\theta = L\int_0^t \frac{dt}{mr^2}$$

$$(\theta)_{\theta_0}^{\theta} = L\int_0^t \frac{dt}{mr^2}$$

$$\theta - \theta_0 = L\int_0^t \frac{dt}{mr^2}$$

$$\theta = L\int_0^t \frac{dt}{mr^2} + \theta_0 \qquad \text{---------------- (25)}$$

Thus ultimately problem has reduce to 2 quadrature's equation (24) & (25) with 4 constants of integration $E, L, r_0 \& \theta_0$. By knowing $r(t) \& \theta(t)$, we can locate position of the particle on the orbital path at any instant 't' and this gives solution of problem.

2.6. CLASSIFICATION OF ORBITS

A system has two generalized co – ordinates r and θ and therefore two second order differential equations have to be solved to study the motion. In other words, four integrations are needed to study the motion of a article. However, without solving the equations of motion for a specific central force, we can learn a lot about the motion using the first integrals of motion, E and L.

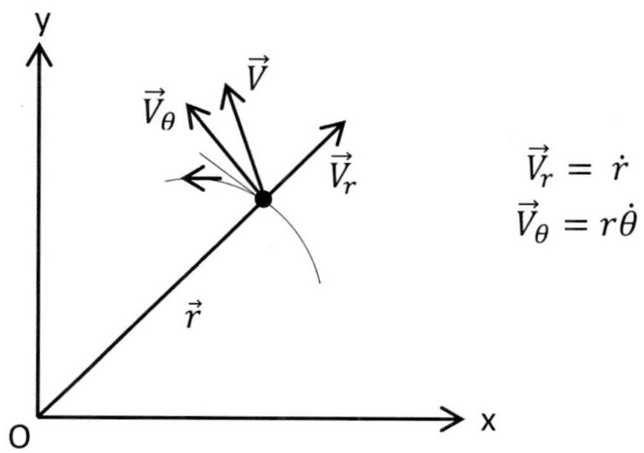

Fig. (2.8)

We know that equation of motion is given by

$$m\ddot{r} - \frac{L^2}{mr^3} - F(r) = 0$$

$$m\ddot{r} = F(r) + \frac{L^2}{mr^3}$$

$$m\ddot{r} = F^{\,|}(r)$$

Let us consider

$$F^{\,|}(r) = F(r) + \frac{L^2}{mr^3}$$

$$F^{\,|}(r) = F(r) + \frac{(mr^2\dot{\theta})^2}{mr^3} \begin{bmatrix} We\ know\ that \\ L = mr^2\dot{\theta} \end{bmatrix}$$

$$F^{\,|}(r) = F(r) + \frac{m^2\, r^4\, \dot{\theta}^2}{mr^3}$$

$$F^{\,|}(r) = F(r) + mr\dot{\theta}^2 \qquad \text{----------------- (1)}$$

Multiply and divide with r to the 2nd term of equation (1), we get

$$F^{\prime}(r) = F(r) + mr\dot{\theta}^2 X \frac{r}{r}$$

$$F^{\prime}(r) = F(r) + \frac{m\, r^2\, \dot{\theta}^2}{r}$$

$$F^{\prime}(r) = F(r) + \frac{m\, (r\, \dot{\theta})^2}{r}$$

$$F^{\prime}(r) = F(r) + \frac{m\, V_\theta{}^2}{r} \qquad \begin{bmatrix} We\ know\ that \\ \vec{V_\theta}^{\,2} = r\dot{\theta} \end{bmatrix}$$

Effective force = Central force + Centrifugal force

Now by definition, we know that

$$F^{\prime}(r) = -\frac{dV^{\prime}}{dr}$$

$$-dV^{\prime} = F^{\prime}(r)\, dr$$

$$-dV^{\prime} = \left[F(r) + \frac{m\, V_\theta{}^2}{r}\right] dr$$

$$-dV^{\prime} = \left[F(r) + \frac{L^2}{mr^3}\right] dr$$

Integrating on both sides, we get

$$-\int dV^{\prime} = \int F(r)\, dr + \int \frac{L^2}{mr^3}\, dr$$

$$\int dV^{\prime} = -\int F(r)\, dr - \frac{L^2}{m} \int \frac{1}{r^3}\, dr$$

$$V^{\prime} = V - \frac{L^2}{m}\left(-\frac{1}{2r^2}\right) \qquad \begin{bmatrix} By\ definition \\ F(r) = -\frac{dV}{dr} \\ dV = -F(r)dr \\ \int dV = -\int F(r)dr \\ V = -\int F(r)dr \end{bmatrix}$$

$$V^{\prime} = V + \frac{L^2}{2mr^2} \qquad \text{----------------- (2)}$$

For energy conservation theorem

$$E = \frac{1}{2}\, m\dot{r}^2 + \frac{L^2}{2mr^2} + V$$

$$E = \frac{1}{2}\, m\dot{r}^2 + V^{\prime}{}_{(r)} \qquad \text{------------ (3)} \quad \text{[(From equation (3)]}$$

Consider the case of an attractive central force obeying inverse square law.

i.e. $\quad F(r) = -\frac{K}{r^2} \qquad (K > 0) \quad \begin{bmatrix} We\ know\ that \\ F(r) = \frac{Gm_1m_2}{r^2} \\ F(r) = -\frac{K}{r^2} \end{bmatrix}$

$$-\frac{dV}{dr} = -\frac{K}{r^2}$$

$$dV = \frac{K}{r^2}\, dr$$

Integrating on both sides, we get

$$\int dV = K \int \frac{1}{r^2}\, dr$$

$$V = -\frac{K}{r} \qquad \text{------------------ (4)}$$

Put value of equation (4) in equation (2), we get

$$V^{\prime} = -\frac{K}{r} + \frac{L^2}{2mr^2}$$

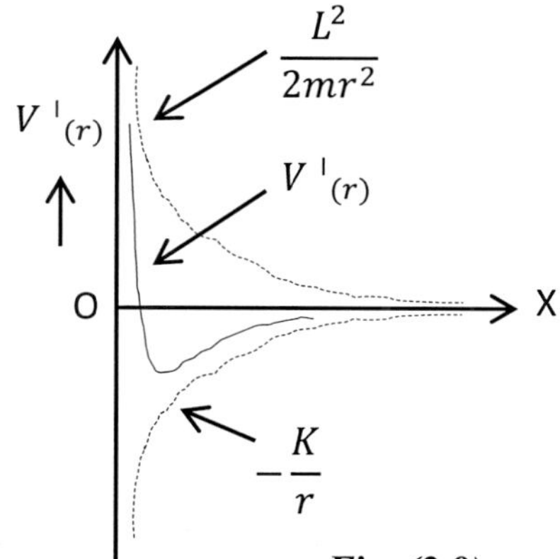

Fig. (2.9)

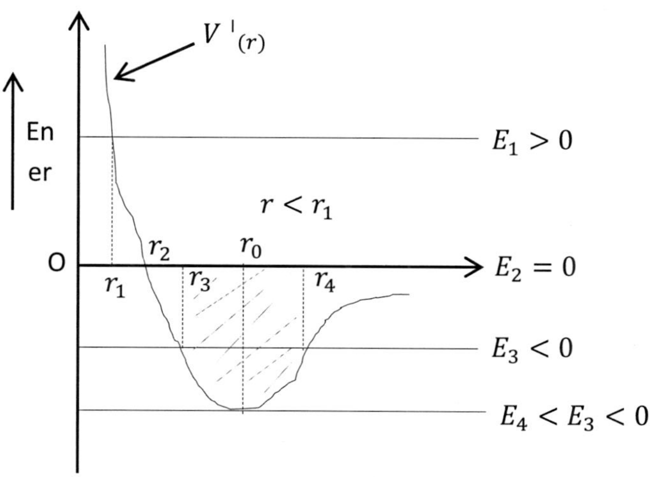

Fig. (2.10)

Case (i):

Let us consider the motion of particle having energy E_1, clearly this particle can never come closer than r_1. If $r < r_1$ $V^{\prime}$ exceeds E_1, then from equation (3) Kinetic energy (K.E) have to be negative (-ve), corresponding to an imaginary velocity which is not possible.

On the other hand there is no upper limit to the possible values of r, so the orbit is not bounded. A particle will come in from infinity (∞), strike the repulsive barrier, be repelled & travel back to infinity as shown below. There is only one turning point at $r = r_1$

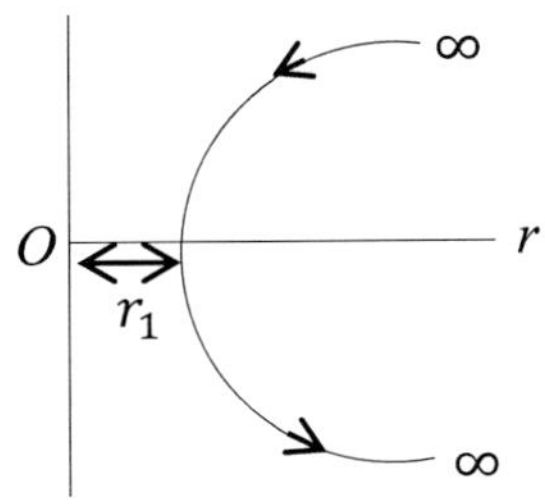

***Fig. (2.11)**: The Hyperbolic orbit for E_1 for bounded motion*

Case (ii):

For energy $E_2 = 0$, a roughly similar picture of the orbit behaviour is obtained.

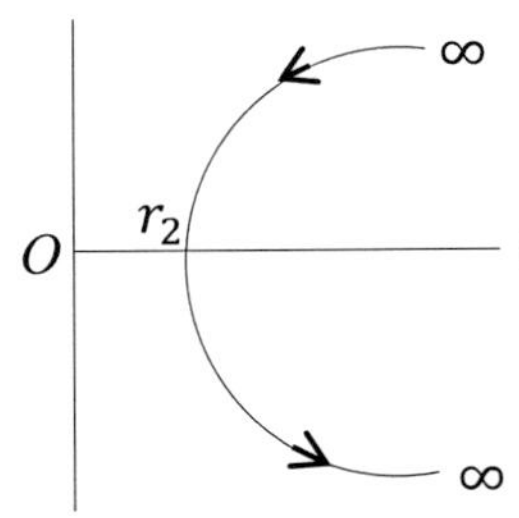

***Fig. (2.12)**: The Parabolic orbit for E_2 for bounded motion*

Case (iii):

Now consider the motion of particle having energy $E_3 < 0$. Here we have a different story.

In this case, motion is bounded with $r_3 \leq r \leq r_4$. The values of r_3 & r_4 are turning points or the apsidal distances of the orbit. This does not necessarily mean that the orbits are closed (elliptical). The motion may be bounded between 2 circles of radius r_3 & r_4 as shown in fig.(2.13)

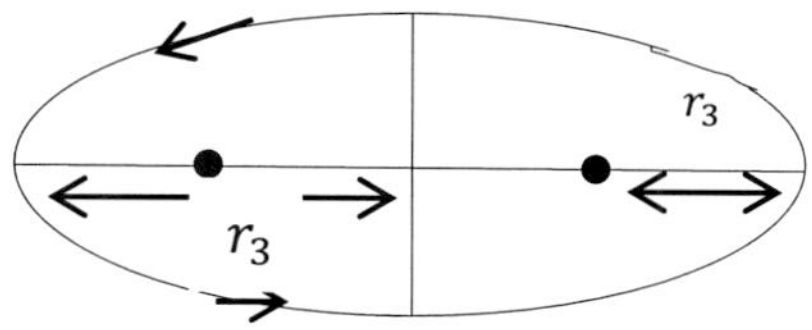

***Fig. (2.12)**: The Elliptical orbit for $E_3 < 0$ for bounded motion*

Case (iv):

Energy E_4 corresponds to minimum value of $V'_{(r)}$

$$V' = -\frac{K}{r} + \frac{L^2}{2mr^2}$$

For V' to be minimum

$$\frac{dV'}{dr} = 0$$

$$\frac{d}{dr}\left[-\frac{K}{r} + \frac{L^2}{2mr^2}\right] = 0$$

$$\frac{K}{r^2} - \frac{L^2}{2mr^3} = 0$$

At $= r_o$, V' is minimum

$$\frac{K}{{r_o}^2} - \frac{L^2}{m{r_0}^3} = 0$$

$$\frac{mkr_0 - L^2}{m{r_o}^3} = 0$$

$$mkr_0 - L^2 = 0\ X\ m{r_o}^3$$

$$mkr_0 - L^2 = 0$$

$$r_0 = \frac{L^2}{mk}$$

At $= r_o$, $E_4 = \left[V'_{(r)}\right]_{mk}$

$$E_4 = -\frac{K}{r} + \frac{L^2}{2mr^2}$$

$$E_4 = -\frac{K\,mk}{L^2} + \frac{L^2\,(mk)^2}{2mL^4}$$

$$E_4 = -\frac{mk^2}{2L^2}$$

In this case 2 bounds r_3 & r_4 coincides i.e. $r_2 = r_3 = r_o$ & motion is possible at only one radius r_o with $\dot{r}_o = 0$. Hence the orbit is a circle & the motion is bounded.

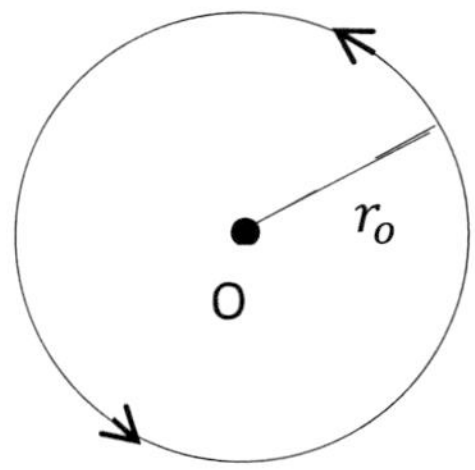

Fig. (2.13): *The Circular orbit for $E_4 < E_3 < 0$ for bounded motion*

2.7. VIRIAL THEOREM

This theorem is valid for large variety of systems is concerned with time averages of various mechanical quantities.

Let us consider a system of N – particles. Let $\vec{r}_i$ be the position vector of ith particle and $\vec{F}_i$ be the force acting on the ith particle. According to Newton's 2nd law of motion, the equation of motion for i^{th} particle is given by

$$\vec{F}_i = \frac{d\vec{P}_i}{dt}$$

$$\vec{F}_i = \dot{\vec{P}}_i \qquad \text{------------------- (1)}$$

Let us introduce a scalar quantity G

Where, $G = \sum_{i=1}^{N} \vec{P}_i.\vec{r}_i$ is the Virial of the given system.

Taking time derivative on both sides.

$$\frac{dG}{dt} = \sum_{i=1}^{N} \vec{P}_i.\dot{\vec{r}}_i + \sum_{i=1}^{N} \dot{\vec{P}}_i.\vec{r}_i$$

$$\frac{dG}{dt} = \sum_{i=1}^{N} m_i\vec{V}_i.\vec{V}_i + \sum_{i=1}^{N} \vec{F}_i.\vec{r}_i \qquad \begin{bmatrix} We\ know\ that \\ \dot{\vec{r}}_i = \frac{d\vec{r}_i}{dt} = \vec{V}_i \\ \vec{P}_i = m_i\vec{V}_i \\ \dot{\vec{P}}_i = \vec{F}_i \end{bmatrix}$$

$$\frac{dG}{dt} = \sum_{i=1}^{N} m_i\vec{V}_i^{\,2} + \sum_{i=1}^{N} \vec{F}_i.\vec{r}_i$$

Now multiply and divide with $1/_2$ to the first term of RHS, we get

$$\frac{dG}{dt} = 2\sum_{i=1}^{N} \frac{1}{2} m_i\vec{V}_i^{\,2} + \sum_{i=1}^{N} \vec{F}_i.\vec{r}_i$$

Where $T = \sum_{i=1}^{N} \frac{1}{2} m_i \vec{V_i}^2$ = Total K.E of system

$$\frac{dG}{dt} = 2\,T + \sum_{i=1}^{N} \vec{F_i}.\vec{r_i} \quad \text{------------------ (2)}$$

The time average of equation (2) over time interval τ is given by

$$\overline{\frac{dG}{dt}} = \frac{1}{\tau}\int_0^{\tau} \frac{dG}{dt} dt = 2\bar{T} + \overline{\sum_{i=1}^{N} \vec{F_i}\,\vec{r_i}} \qquad \left[\begin{array}{c} We\ know\ that \\ \overline{f(t)} = \frac{1}{\tau}\int_0^{\tau} f(t)\,dt \end{array}\right]$$

$$\frac{1}{\tau}\,[G]_0^{\tau} = 2\bar{T} + \overline{\sum_{i=1}^{N} \vec{F_i}\,\vec{r_i}}$$

$$\frac{1}{\tau}\,[F(\tau) - G(0)] = 2\bar{T} + \overline{\sum_{i=1}^{N} \vec{F_i}\,\vec{r_i}} \quad \text{--------------- (3)}$$

Case (i):
If the motion is produce i.e. all co-ordinates repeat after some time and τ is choosen to be the time period then $G(\tau) = G(0)$ then from equation (3), we get

$$0 = 2\bar{T} + \overline{\sum_{i=1}^{N} \vec{F_i}\,\vec{r_i}}$$

$$\bar{T} = -\frac{1}{2}\,\overline{\sum_{i=1}^{N} \vec{F_i}\,\vec{r_i}}$$

Case (ii):
If the motion is non – periodic such that $G(\tau) \neq G(0)$ then LHS of equation (3) can be made zero by choosing τ efficiently large i.e. $\lim_{\tau \to 0} \frac{1}{\tau}\,[G(\tau) - G(0)] \;\to 0$ then from equation (3), we have

$$0 = 2\bar{T} + \overline{\sum_{i=1}^{N} \vec{F_i}\,\vec{r_i}}$$

$$\bar{T} = -\frac{1}{2}\,\overline{\sum_{i=1}^{N} \vec{F_i}\,\vec{r_i}} \quad \text{------------------- (4)}$$

Equation (4) is known as Virial theorem and $-\frac{1}{2}\,\overline{\sum_{i=1}^{N} \vec{F_i}\,\vec{r_i}}$ is known as Virial of clausius.

In case, if i^{th} particle is moving under central force field $\vec{F_i}$ then by definition

$$\vec{F_i} = -\frac{\partial V}{\partial \vec{r_i}} \quad \text{------------------- (5)}$$

Put $\vec{F_i}$ value in equation (4), we get

$$\bar{T} = -\frac{1}{2}\overline{\sum_{i=1}^{N}\left(-\frac{\partial V}{\partial \vec{r_i}}\right)\vec{r_i}}$$

$$\bar{T} = \frac{1}{2}\overline{\sum_{i=1}^{N}\frac{\partial V}{\partial \vec{r_i}}\vec{r_i}} \text{ ------------------ (6)}$$

Let $V = C\,r^k$

Taking partial differential on both sides, we get

$$\frac{\partial V}{\partial r} = Ckr^{k-1}$$

Multiply with r on both sides, we get

$$\frac{\partial V}{\partial r}.r = Ckr^{k-1}.r$$

$$\frac{\partial V}{\partial r}.r = k(Cr^k)$$

$$\frac{\partial V}{\partial r}.r = kV \text{ --------------- (7)} \qquad \begin{bmatrix} We\ know\ that \\ V = Cr^k \end{bmatrix}$$

Put equation (7) value in equation (6), we get

$$\bar{T} = \frac{1}{2}k\,\bar{V} \text{ ------------------- (8)}$$

We know that

$$E = T + V$$

The time average of this equation can be written as

$$\bar{E} = \bar{T} + \bar{V}$$

$$E = \bar{T} + \bar{V}$$

$$E = \frac{1}{2}k\,\bar{V} + \bar{V}$$

$$E = \left(\frac{1}{2}k + 1\right)\bar{V}$$

$$\bar{V} = \frac{2E}{k+2} \text{ ------------------ (9)}$$

Put $\bar{V}$ value in equation (8), we get

$$\bar{T} = \frac{1}{2}k\left[\frac{2E}{k+2}\right]$$

$$\bar{T} = \frac{kE}{k+2}$$

For k = 1

$$\bar{T} = \frac{-E}{1} \qquad \text{(OR)} \ -\bar{T} = E$$

2.8. LAPLACE RUNGE – LENZ VECTOR

The Kepler's problem is also distinguished by the existence of an additional conserved besides the angular momentum. Let us consider a particle of mas 'm' moving in a central force field $\vec{F}$ is given by

$$\vec{F} = f(r)\hat{r}$$

$$\vec{F} = f(r)\frac{\vec{r}}{r} \quad \text{---------------} \ (1)$$

According to Newton's 2nd law of motion

$$\frac{d\vec{P}}{dt} = \vec{F} \quad \text{---------------} \ (2)$$

Comparing equation (1) and (2), we get

$$\frac{d\vec{P}}{dt} = f(r).\frac{\vec{r}}{r} \quad \text{---------------} \ (3)$$

Consider a vector $\vec{R}$ such that

$$\vec{R} = \vec{P} \, X \, \vec{L}$$

Differentiating on both sides, w.r.to. 't', we get

$$\frac{d\vec{R}}{dt} = \vec{P} \, X \, \frac{d\vec{L}}{dt} + \frac{d\vec{P}}{dt} \, X \, \vec{L}$$

$$\frac{d\vec{R}}{dt} = \vec{P} \, X \, 0 + \frac{d\vec{P}}{dt} \, X \, \vec{L} \qquad \begin{bmatrix} We\ know\ that \\ \vec{L}\ is\ constant \\ So\ \frac{d\vec{L}}{dt} = 0 \end{bmatrix}$$

$$\frac{d\vec{R}}{dt} = 0 + \frac{d\vec{P}}{dt} \, X \, \vec{L}$$

$$\frac{d\vec{R}}{dt} = \frac{d\vec{P}}{dt} \, X \, \vec{L} \quad \text{-----------------} \ (4)$$

Put equation (3) value in equation (4), we get

$$\frac{d\vec{R}}{dt} = f(r).\frac{\vec{r}}{r} \, X \, \vec{L}$$

$$\frac{d\vec{R}}{dt} = f(r).\frac{\vec{r}}{r}\ X\left(\vec{r}\ X\ \vec{P}\right) \qquad \begin{bmatrix} We\ know\ that \\ \vec{L} = \vec{r}\ X\ \vec{P} \end{bmatrix}$$

$$\frac{d\vec{R}}{dt} = f(r).\frac{\vec{r}}{r}\ X\left(\vec{r}\ X\ m\vec{V}\right) \qquad \begin{bmatrix} We\ know\ that \\ \vec{P} = m\vec{V} \end{bmatrix}$$

$$\frac{d\vec{R}}{dt} = \frac{mf(r)}{r}\left[\vec{r}X\left(\vec{r}\ X\ \dot{\vec{r}}\right)\right] \qquad \begin{bmatrix} We\ know\ that \\ \vec{V} = \frac{d\vec{r}}{dt} = \dot{\vec{r}} \end{bmatrix}$$

$$\frac{d\vec{R}}{dt} = \frac{mf(r)}{r}\left[(\vec{r}.\dot{\vec{r}})\vec{r} - (\vec{r}.\vec{r})\dot{\vec{r}}\right] \qquad \begin{bmatrix} We\ know\ that \\ Vector\ product \\ \vec{A}\ X\ (\vec{B}\ X\ \vec{C}) = (\vec{A}.\vec{C})\vec{B} - (\vec{A}.\vec{B})\vec{C} \end{bmatrix}$$

$$\frac{d\vec{R}}{dt} = \frac{mf(r)}{r}\left[(r.\dot{r})\vec{r} - (r^2)\dot{\vec{r}}\right] \qquad \begin{bmatrix} We\ know\ that \\ \vec{r}.\dot{\vec{r}} = r.\dot{r} \\ \vec{r}.\vec{r} = r.r = r^2 \end{bmatrix}$$

$$\frac{d\vec{R}}{dt} = -\frac{mf(r)}{r}\left[(r^2)\dot{\vec{r}} - (r.\dot{r})\vec{r}\right]$$

$$\frac{d\vec{R}}{dt} = -\frac{mf(r)}{r}\ r^2\left[\dot{\vec{r}} - \frac{r\dot{r}\vec{r}}{r^2}\right]$$

$$\frac{d\vec{R}}{dt} = -mf(r).r^2\left[\frac{\dot{\vec{r}}}{r} - \frac{r\dot{r}\vec{r}}{r^3}\right]$$

$$\frac{d\vec{R}}{dt} = -mf(r).r^2\left[\frac{\dot{\vec{r}}}{r} - \frac{\dot{r}\vec{r}}{r^2}\right]$$

$$\frac{d\vec{R}}{dt} = -mf(r).r^2\left[\frac{d}{dt}\left(\frac{\vec{r}}{r}\right)\right] \text{ --------- (5)} \begin{bmatrix} \frac{d}{dt}\left(\vec{r}.\frac{1}{r}\right) = \frac{1}{r}\dot{\vec{r}} + \vec{r}\left(-\frac{1}{r^2}\right)\dot{r} \\ \frac{d}{dt}\left(\frac{\vec{r}}{r}\right) = \frac{\dot{\vec{r}}}{r} - \frac{\dot{r}\vec{r}}{r^2} \end{bmatrix}$$

For central force $f(r) = \pm\frac{K}{r^2}$ ----------- (6)

$$\frac{d\vec{R}}{dt} = -m\left(\pm\frac{K}{r^2}\right).r^2\left[\frac{d}{dt}\left(\frac{\vec{r}}{r}\right)\right]$$

$$\frac{d\vec{R}}{dt} = \mp mk\ \frac{d}{dt}\left(\frac{\vec{r}}{r}\right)$$

$$\frac{d\vec{R}}{dt} \pm mk\ \frac{d}{dt}\left(\frac{\vec{r}}{r}\right) = 0$$

$$\frac{d}{dt}\left[\vec{R} \pm mk\frac{\vec{r}}{r}\right] = 0$$

$$\frac{d}{dt}\left[\vec{P}X\ \vec{L} \pm mk\frac{\vec{r}}{r}\right] = 0$$

$$\vec{P} X \vec{L} \pm mk\frac{\vec{r}}{r} = Constant$$

This is known as Laplace – Runge Lenz Vector ($\vec{A}$)

i.e. $\vec{A}$ = $\vec{P} X \vec{L} \pm mk\frac{\vec{r}}{r}$ = Constant

Multiply $\vec{L}$ on both sides, we get

$$\vec{A} \text{ x } \vec{L} = (\vec{P} X \vec{L}).\vec{L} \pm mk\frac{\vec{r}}{r}.\vec{L}$$

$$\vec{A} \text{ x } \vec{L} = 0 \pm \frac{mk}{r}.\vec{r}(\vec{r} X \vec{P}) \qquad \begin{bmatrix} We\ know\ that \\ (\vec{A} X \vec{B}).\vec{B} = 0 \end{bmatrix}$$

$$\vec{A} \text{ x } \vec{L} = 0 \pm 0$$

$$\vec{A} \text{ x } \vec{L} = 0$$

i.e. $\vec{A} \perp \vec{L}$

As $\vec{L}$ is perpendicular to the plane of motion so $\vec{A}$ will always be in plane of motion.

2.9. THE DIFFERENTIAL EQUATION FOR THE ORBIT AND INTEGRABLE POWER LAW POTENTIAL

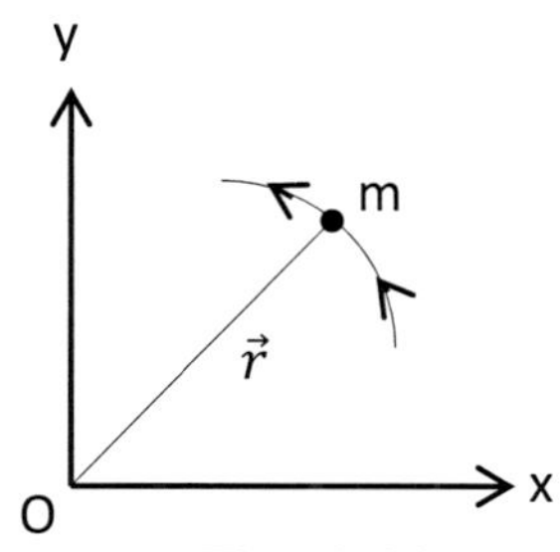

Fig. (2.14)

The equation of motion for a particle of reduced mass 'm' moving under central force of magnitude f(r) is given by

$$m\ddot{r} - \frac{L^2}{mr^3} = f(r) \quad \text{-------------- (1)}$$

Now $\dot{r} = \frac{dr}{dt}$

Multiply and divide with , we get

$$\dot{r} = \frac{dr}{d\theta}.\frac{d\theta}{dt}$$

$$\dot{r} = \frac{dr}{d\theta}.\dot{\theta} \qquad \text{---------------- (2)}$$

As we know $L = mr^2\dot{\theta}$

$$\dot{\theta} = \frac{L}{mr^2} \qquad \text{------------- (3)}$$

Put $\dot{\theta}$ value in equation (2), we get

$$\dot{r} = \frac{dr}{d\theta}.\left(\frac{L}{mr^2}\right) \qquad \text{------------- (4)}$$

Now again differentiate w.r.to time 't' i.e. time derivative of equation (4) is given by

$$\ddot{r} = \frac{d}{dt}\left[\frac{L}{mr^2}.\frac{dr}{d\theta}\right]$$

Multiply and divide with $d\theta$ on RHS side, we get

$$\ddot{r} = \frac{d}{dt}\left[\frac{L}{mr^2}.\frac{dr}{d\theta}\right].\frac{d\theta}{d\theta}$$

$$\ddot{r} = \frac{d}{d\theta}\left[\frac{L}{mr^2}.\frac{dr}{d\theta}\right].\frac{d\theta}{dt}$$

$$\ddot{r} = \frac{d}{d\theta}\left[\frac{L}{mr^2}.\frac{dr}{d\theta}\right].\dot{\theta}$$

$$\ddot{r} = \frac{d}{d\theta}\left[\frac{L}{mr^2}.\frac{dr}{d\theta}\right].\frac{L}{mr^2}$$

$$\ddot{r} = \frac{L}{mr^2}\left\{\frac{d}{d\theta}\left[\frac{L}{mr^2}.\frac{dr}{d\theta}\right]\right\} \qquad \text{-------------- (5)}$$

Let $u = \frac{1}{r}$ ------------------- (6)

$$\frac{du}{d\theta} = -\frac{1}{r^2}\frac{dr}{d\theta}$$

$$\frac{du}{d\theta} = -u^2\frac{dr}{d\theta} \qquad \begin{bmatrix}\text{We know that}\\ u = \frac{1}{r}\end{bmatrix}$$

$$\frac{dr}{d\theta} = -\frac{1}{u^2}\frac{du}{d\theta} \qquad \text{------------- (7)}$$

Put equation (6) value & equation (7) value in equation (5), we get

$$\ddot{r} = \frac{Lu^2}{m}\left\{\frac{d}{d\theta}\left[\frac{Lu^2}{m}.-\frac{1}{u^2}\frac{du}{d\theta}\right]\right\}$$

$$\ddot{r} = -\frac{Lu^2}{m^2}\frac{d}{d\theta}\left(\frac{du}{d\theta}\right)$$

$$\ddot{r} = -\frac{Lu^2}{m^2}\frac{d^2u}{d\theta^2} \qquad \text{----------------- (8)}$$

Put equation (8) value and equation (6) value in equation (1), we get

$$m\left[-\frac{Lu^2}{m^2}\frac{d^2u}{d\theta^2}\right] - \frac{L^2u^3}{m} = f\left(\frac{1}{u}\right)$$

$$-\frac{L^2u^2}{m}\left[\frac{d^2u}{d\theta^2} + u\right] = f\left(\frac{1}{u}\right)$$

$$\frac{d^2u}{d\theta^2} + u = -\frac{m}{L^2u^2} f\left(\frac{1}{u}\right) \qquad \text{-------------------- (9)}$$

By definition central force can be written as

$$f(r) = -\frac{\partial V_{(r)}}{\partial r}$$

Now, multiply & divide with ∂u on RHS, we get

$$f(r) = -\frac{\partial V_{(r)}}{\partial u} \cdot \frac{\partial u}{\partial r} \qquad \text{-------------------- (10)}$$

As $\qquad u = \frac{1}{r}$

So $\qquad \frac{\partial u}{\partial r} = -\frac{1}{r^2} = -u^2$

$$\frac{\partial u}{\partial r} = -u^2 \qquad \text{------------------- (11)}$$

Put the value of equation (11) in equation (10), we get

$$f\left(\frac{1}{u}\right) = -\frac{\partial V_{(1/u)}}{\partial u} \cdot (-u^2)$$

$$f\left(\frac{1}{u}\right) = u^2 \frac{\partial V_{(1/u)}}{\partial u} \qquad \text{----------------- (12)}$$

Put the value of equation (12) in equation (9), we get
Equation (9) becomes

$$\frac{d^2u}{d\theta^2} + u = -\frac{m}{L^2u^2} f\left(\frac{1}{u}\right)$$

$$\frac{d^2u}{d\theta^2} + u = -\frac{m}{L^2u^2}\left[u^2\frac{\partial V(^1/_u)}{\partial u}\right]$$

$$\frac{d^2u}{d\theta^2} + u = -\frac{m}{L^2}\left[\frac{\partial V(^1/_u)}{\partial u}\right]$$

This is the required differential equation for the orbit we know that

$$dt = \frac{dr}{\sqrt{\frac{2}{m}\left(E - V - \frac{L^2}{2mr^2}\right)}} \quad \text{----------} \ (13) \quad \begin{pmatrix} \textit{From equations} \\ \textit{of motion and} \\ \textit{first integrals} \end{pmatrix}$$

As $\quad \dot{\theta} = \frac{L}{mr^2}$

$$\frac{d\theta}{dt} = \frac{L}{mr^2}$$

$$dt = \frac{mr^2}{L} d\theta \quad \text{---------------} \ (14)$$

Put the value of equation (14) in equation (13), we get

$$\frac{mr^2}{L} d\theta = \frac{dr}{\sqrt{\frac{2}{m}\left(E - V - \frac{L^2}{2mr^2}\right)}}$$

$$d\theta = \frac{L}{mr^2} \frac{dr}{\sqrt{\frac{2}{m}\left(E - V - \frac{L^2}{2mr^2}\right)}}$$

$$d\theta = \frac{1}{\frac{mr^2}{L}} \frac{dr}{\sqrt{\frac{2}{m}\left(E - V - \frac{L^2}{2mr^2}\right)}}$$

$$d\theta = \frac{1}{r^2} \frac{dr}{\sqrt{\frac{2}{m} X \frac{m^2}{L^2}\left(E - V - \frac{L^2}{2mr^2}\right)}}$$

$$d\theta = \frac{1}{r^2} \frac{dr}{\sqrt{\frac{2m}{L^2}\left(E - V - \frac{L^2}{2mr^2}\right)}}$$

$$d\theta = \frac{dr}{r^2 . \sqrt{\frac{2m}{L^2}\left(E - V - \frac{L^2}{2mr^2}\right)}}$$

$$d\theta = \frac{dr}{r^2 . \sqrt{\frac{2mE}{L^2} - \frac{2mV}{L^2} - \frac{L^2}{2mr^2} X \frac{2m}{L^2}}}$$

$$d\theta = \frac{dr}{r^2 \cdot \sqrt{\frac{2mE}{L^2} - \frac{2mV}{L^2} - \frac{1}{r^2}}}$$

Put $\quad r = \frac{1}{u}$

$$dr = -\frac{1}{u^2} du$$

So $\quad d\theta = \frac{u^2\left(-\frac{1}{u^2} du\right)}{r^2 \cdot \sqrt{\frac{2mE}{L^2} - \frac{2mV}{L^2} - u^2}}$

$$d\theta = \frac{-du}{r^2 \cdot \sqrt{\frac{2mE}{L^2} - \frac{2mV}{L^2} - u^2}}$$

Integrating on both sides with limits $\theta_0 \; to \; \theta$ & $u_0 \; to \; u$, we get

$$\int_{\theta_0}^{\theta} d\theta = -\int_{u_0}^{u} \frac{du}{\sqrt{\frac{2mE}{L^2} - \frac{2mV}{L^2} - u^2}}$$

$$(\theta)_{\theta_0}^{\theta} = -\int_{u_0}^{u} \frac{du}{\sqrt{\frac{2mE}{L^2} - \frac{2mV}{L^2} - u^2}}$$

$$\theta - \theta_0 = -\int_{u_0}^{u} \frac{du}{\sqrt{\frac{2mE}{L^2} - \frac{2mV}{L^2} - u^2}}$$

$$\theta = \theta_0 - \int_{u_0}^{u} \frac{du}{\sqrt{\frac{2mE}{L^2} - \frac{2mV}{L^2} - u^2}} \qquad \text{-------------- (a)}$$

Let $\quad V = kr^{n+1}$ --------------- (b)

As $\quad f = -\frac{\partial V}{\partial r}$

$$f = -\frac{\partial\,(kr^{n+1})}{\partial r} \qquad [Put\;\; V = kr^{n+1}]$$

$$f = -k(n+1)r^n$$

i.e. Central force varies as n[th] power of '*r*'

Put the value of equation (b) in equation (a), we get

$$\theta = \theta_0 - \int_{u_0}^{u} \frac{du}{\sqrt{\frac{2mE}{L^2} - \frac{2mkr^{n+1}}{L^2} - u^2}}$$

$$\theta = \theta_0 - \int_{u_0}^{u} \frac{du}{\sqrt{\frac{2mE}{L^2} - \frac{2mk\left(\frac{1}{u}\right)^{n+1}}{L^2} - u^2}} \quad \left[Put\ r = \frac{1}{u}\right]$$

$$\theta = \theta_0 - \int_{u_0}^{u} \frac{du}{\sqrt{\frac{2mE}{L^2} - \frac{2mku^{-(n+1)}}{L^2} - u^2}} \quad \text{---------- (c)}$$

Let $V = kr^{n+1}$ --------- (b) $f = k(n+1)r^n$

To find the value of integral (c) it is convenient to reduce the denominator in the form $\sqrt{\alpha u^2 + \beta u + r}$ for which $-(n+1) = 0,1,\ldots..$

Case (i): If $-(n+1) = 0$

Then $n = -1$

$\therefore$ $V = kr^0$

$V = K\ (Constant)$

i.e. Particle is moving under constant potential. So it is not possible because potential depends upon r in a central force field.

Case (ii): If $-(n+1) = 1$

Then $-n - 1 = 1$

$n = -2$

$f \propto \frac{1}{r^2}$

This showing Inverse square law

Here force is varying with $\frac{1}{r^2}$

Case (iii): If $-(n+1) = 2$

Then $n = -3$ & So on

$f \propto \frac{1}{r^3}$

This showing Inverse cube law

Here force is varying with $\frac{1}{r^3}$

2.10. THE KEPLER PROBLEM: INVERSE SQUARE LAW OF FORCE

The inverse square law is most important of all the central force laws. It results the deduction of Kepler's laws of planetary motion.

2.10.1 KEPLLER'S FIRST LAW OF PLANETORY MOTION (OR) THE LAW OF ELLIPTICAL ORBITS

Statement:

All the planets moving around the sun in an elliptical orbit with the sun is one of the foci.

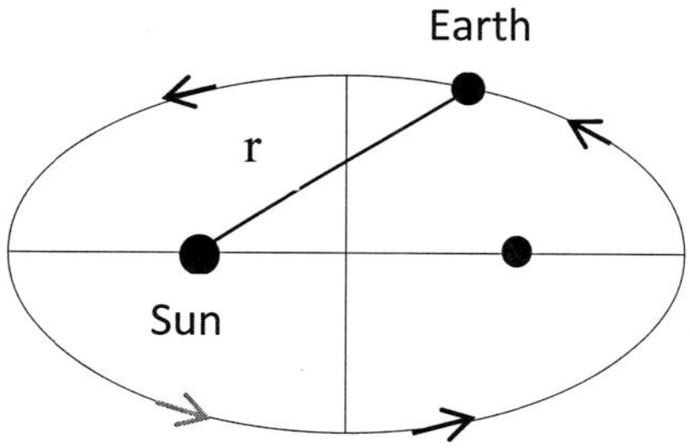

Fig. (2.15)

Proof:

We know that differential equation of the orbit is given by

$$\frac{d^2u}{d\theta^2} + u = -\frac{m}{L^2u^2} f\left(\frac{1}{u}\right) \text{ ----------------- (1)}$$

For central force obeying inverse square law

$$F(r) = -\frac{k}{r^2}$$

Put $\quad r = \frac{1}{u} \quad$ we get

$$f\left(\frac{1}{u}\right) = -ku^2 \text{ ----------------- (2)}$$

Put the value of equation (2) in equation (1), we get

$$\frac{d^2u}{d\theta^2} + u = -\frac{m}{L^2u^2}(-ku^2)$$

$$\frac{d^2u}{d\theta^2} + u = \frac{mk}{L^2}$$

$$\frac{d^2u}{d\theta^2} + u - \frac{mk}{L^2} = 0 \qquad \text{------------------ (3)}$$

Let $$u - \frac{mk}{L^2} = x \qquad \text{------------------ (4)}$$

Differentiate with respect to θ, we get

$$\frac{du}{d\theta} = \frac{dx}{d\theta}$$

Now apply squaring on both sides, we get

$$\frac{d^2u}{d\theta^2} = \frac{d^2x}{d\theta^2} \qquad \text{------------------ (5)}$$

Put the value of equation (4) & (5) in equation (3), we get

$$\frac{d^2x}{d\theta^2} + x = 0 \qquad \text{------------------ (6)}$$

The general equation for equation (6) is given by

$$x = A\,Cos(\theta - \theta^{\,\prime}) \qquad \text{------------------ (7)}$$

Where A & $\theta^{\,\prime}$ are constants of integration, comparing equation (4) & (7) we get

$$u - \frac{mk}{L^2} = A\,Cos(\theta - \theta^{\,\prime}) \qquad \text{---------------- (8)}$$

$$u = \frac{mk}{L^2} + A\,Cos(\theta - \theta^{\,\prime})$$

$$\frac{1}{r} = \frac{mk}{L^2} + A\,Cos(\theta - \theta^{\,\prime}) \qquad \text{---------------- (9)}$$

Multiplying both side by $\frac{L^2}{mk}$, we get

$$\frac{L^2/mk}{r} = \frac{mk}{L^2}\,X\,\frac{L^2}{mk} + \frac{L^2}{mk}A\,Cos(\theta - \theta^{\,\prime})$$

$$\frac{L^2/mk}{r} = 1 + \frac{L^2}{mk}.A\,Cos(\theta - \theta^{\,\prime}) \qquad \text{------------ (10)}$$

Comparing equation (10) with general equation of conic section

$\frac{l}{r} = 1 + e\,Cos(\theta - \theta^{\,\prime})$ we get

$$l = \frac{L^2}{mk} \quad \& \quad e = \frac{L^2A}{mk}$$

Differentiating equation (9) on both sides, we get

$$-\frac{1}{r^2}\dot{r} = 0 - A\,Sin(\theta - \theta^{\,|})\dot{\theta}$$

$$-\frac{1}{r^2}\dot{r} = -A\,Sin(\theta - \theta^{\,|})\dot{\theta}$$

$$\dot{r} = r^2.A\,Sin(\theta - \theta^{\,|})\dot{\theta} \qquad \text{---------------- (11)}$$

$$L = mr^2\dot{\theta}$$

$$\dot{\theta} = \frac{L}{mr^2} \qquad \text{---------------- (12)}$$

Put the value of equation (12) in equation (11), we get

$$\dot{r} = r^2.A\,Sin(\theta - \theta^{\,|})\,\frac{L}{mr^2}$$

$$\dot{r} = \frac{AL}{m}\,Sin(\theta - \theta^{\,|}) \qquad \text{-------------- (13)}$$

Now total energy of a particle moving under a central force field is given by

$$E = \frac{1}{2}m\dot{r}^2 + V(r) + \frac{L^2}{2mr^2}$$

$$E = \frac{1}{2}m\dot{r}^2 - \frac{k}{r} + \frac{L^2}{2mr^2} \qquad \text{------- (14)} \quad \left[V(r) = -\frac{k}{r}\right]$$

Put the value of equation (9) & (13) in equation (14), we get

$$E = \frac{1}{2}m\left[\frac{AL}{m}\,Sin(\theta - \theta^{\,|})\right]^2 - k\left[\frac{mk}{L^2} + A\,Cos(\theta - \theta^{\,|})\right] + \frac{L^2}{2m}\left[\frac{mk}{L^2} + A\,Cos(\theta - \theta^{\,|})\right]^2$$

$$E = \frac{A^2L^2}{2m} - \frac{mk^2}{2L^2}$$

$$E + \frac{mk^2}{2L^2} = \frac{A^2L^2}{2m}$$

$$A^2 = \frac{2m}{L^2}\left[E + \frac{mk^2}{2L^2}\right]$$

$$A^2 = \frac{2m}{L^2}\;X\;\frac{mk^2}{2L^2}\left[\frac{E}{\frac{mk^2}{2L^2}} + 1\right]$$

$$A^2 = \frac{2m^2k^2}{L^4}\left[1 + \frac{2L^2E}{mk^2}\right]$$

Squaring on both sides we get

$$A = \frac{mk}{L^2}\sqrt{\left(1 + \frac{2L^2E}{mk^2}\right)}$$

As $\quad e = \frac{L^2A}{mk}$

So $\quad e = \frac{L^2}{mk}\left[\frac{mk}{L^2}\sqrt{\left(1 + \frac{2L^2E}{mk^2}\right)}\right]$

$$e = \sqrt{\left(1 + \frac{2L^2E}{mk^2}\right)}$$

Squaring on both sides, we get

$$e^2 = \left[\sqrt{\left(1 + \frac{2L^2E}{mk^2}\right)}\right]^2$$

$$e^2 = 1 + \frac{2L^2E}{mk^2}$$

$$e^2 - 1 = \frac{2L^2E}{mk^2}$$

$$E = \frac{mk^2}{2L^2}(e^2 - 1)$$

The magnitude of 'e' decides the nature of orbit.

Value of 'e'	**Value of total energy 'E'**	**Nature of orbit**
e > 1	E > 0	Hyperbola
e = 1	E = 0	Parabola
e < 1	E < 0	Ellipse
e = 0	E = $-\frac{mk^2}{2L^2}$	Circle

From Kepler's first law of planetary motion we know that when e < 1 & E < 0, then orbit is elliptical & equation of the orbit is given by

$$\frac{l}{r} = 1 + e\ Cos(\theta - \theta^{\prime}) \text{ --------------- } (15)$$

Where

$$l = \frac{L^2}{mk} \text{ ------------ } (16)$$

&

$$e = \sqrt{1 + \frac{2L^2E}{mk^2}} \text{ ------------ } (17)$$

When $\theta - \theta^{\prime} = 0$ Then $Cos(\theta - \theta^{\prime}) = 1\ (Maximum)$
In this case r will be minimum

Let $r_{min} = r_1 = \begin{pmatrix} Perihelion = The\ point\ A\ at \\ which\ earth\ closest\ to\ sun \end{pmatrix}$

$$\frac{l}{r_1} = 1 + e \qquad \frac{l}{r_2} = 1 - e$$

$$r_1 = \frac{l}{1+e} \text{ ------ } (18) \qquad r_2 = \frac{l}{1-e} \text{ ------- } (19)$$

From Figure (2.17), we have

$$2a = r_1 + r_2 \text{ ------------- } (20)$$

Put the values of equation (18) & (19) in equation (20), we get

$$2a = \frac{l}{1+e} + \frac{l}{1-e}$$

$$2a = l\left[\frac{1}{1+e} + \frac{1}{1-e}\right]$$

$$2a = l\left[\frac{2}{1-e^2}\right]$$

$$a = \frac{l}{2}\left[\frac{2}{1-e^2}\right]$$

$$a = \frac{l}{1-e^2} \text{ ------------- } (21)$$

Put the value of equation (16) & (17) in equation (21), we get

$$a = \frac{\frac{L^2}{mk}}{1-\left(\sqrt{1+\frac{2L^2E}{mk^2}}\right)^2}$$

$$a = \frac{\frac{L^2}{mk}}{1-1-\frac{2L^2E}{mk^2}}$$

$$a = \frac{L^2}{mk} X \frac{mk^2}{-2EL^2}$$

$$a = \frac{-k}{2E}$$

$$E = \frac{-k}{2a} \quad \text{------------------ (22)}$$

2.10.2 KEPLLER'S SECOND LAW OF PLANETORY MOTION (AREAL VELOCITY)

Statement:
The line joining the planet and sun sweeps equal areas in equal areas in equal intervals of time.

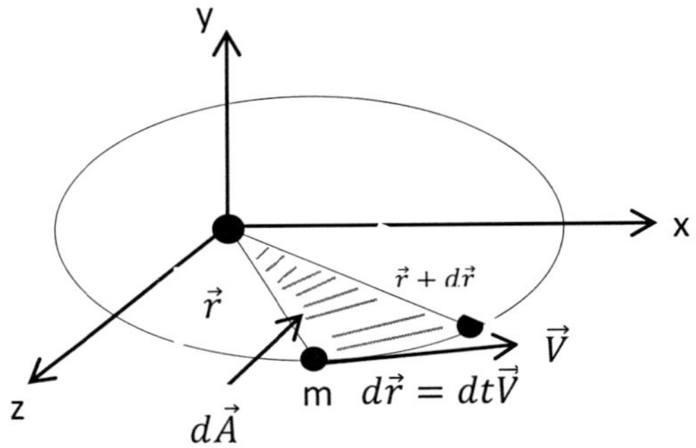

Fig. (2.16)

Kepler's second law of planetary motion describes the speed of planet travelling in an elliptical orbit around the sun. It states that a line between the sun and the planet seeps equal areas in equal interval of times. Thus the speed of the planet increases as it nears the sun and decreases as it recedes from the sun.

Proof:
We know that $d\vec{A} = \frac{1}{2}(\vec{r} X d\vec{r})$
Dividing on both sides by dt, we get

$$\frac{d\vec{A}}{dt} = \frac{1}{2}\left(\vec{r} X \frac{d\vec{r}}{dt}\right)$$

$$\frac{d\vec{A}}{dt} = \frac{1}{2}(\vec{r} X \vec{V})$$

Multiply and divide with 'm' to RHS, we get

$$\frac{d\vec{A}}{dt} = \frac{1}{2m}\left(\vec{r}\, X\, m\vec{V}\right)$$

$$\frac{d\vec{A}}{dt} = \frac{\vec{L}}{2m} \qquad \left[\vec{L} = \vec{r}\, X\, m\vec{V}\right]$$

$$\frac{d\vec{A}}{dt} = Constant$$

Here $\quad \vec{A} \rightarrow Areal\ velocity$
$\quad A \rightarrow is\ constant$

2.10.3 KEPLLER'S THIRD LAW OF PLANETORY MOTION (OR) THE LAW OF PERIODS OR HARMONIC LAW

Statement:

The square of the time period of revolution of a planet around the Sun is directly proportional to the cube of the sum of major axis of the ellipse.

$$\therefore\ T^e \propto a^3$$

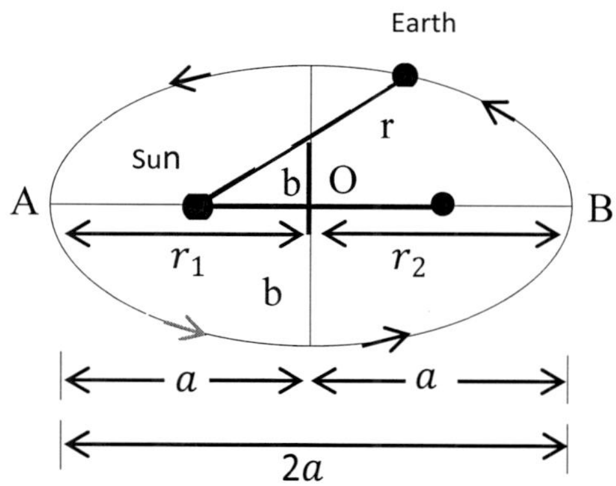

Fig. (2.17)

Proof:

Thus in case of elliptical orbits, the total energy of planet depends only on the major axis($2a$). Let T be the time period of revolution of planet around the sun.

Area of orbit $\quad A = \int_0^T \frac{dA}{dt} dt$

$$A = \int_0^T \frac{1}{2} r^2 \dot{\theta}\, dt \quad \text{------------ (1)}$$

$$A = \int_0^T \frac{1}{2} r^2 \frac{L}{mr^2} dt \qquad \left[\dot{\theta} = \frac{L}{mr^2}\right]$$

$$A = \frac{L}{2m}\int_0^T dt$$

$$A = \frac{L}{2m}\,(t)_0^T$$

$$A = \frac{L}{2m}\,(T-0)$$

$$A = \frac{LT}{2m} \qquad \text{---------------} \ (2)$$

But area of ellipse $\quad A = \pi ab \quad$ --------------- (3)

Comparing equations (2) & (3) we get

$$\frac{LT}{2m} = \pi ab$$

$$T = \frac{2m\pi ab}{L} \qquad \text{--------------} \ (4)$$

As $\quad b = a\sqrt{1-e^2}$

$$b = a\sqrt{1-\left(\sqrt{1+\frac{2L^2E}{mk^2}}\right)^2} \qquad \left[\begin{matrix} We\ know\ that \\ e = \sqrt{\left(1+\frac{2L^2E}{mk^2}\right)} \end{matrix}\right]$$

$$b = a\sqrt{1-\ 1-\frac{2L^2E}{mk^2}}$$

$$b = a\sqrt{-\frac{2L^2}{mk^2}\ X\ \frac{-k}{2a}} \qquad \left[\begin{matrix} We\ know\ that \\ E = \frac{-k}{2a} \end{matrix}\right]$$

$$b = a\sqrt{\frac{L^2}{mka}}$$

$$b = \frac{a}{\sqrt{a}}\cdot\frac{L}{\sqrt{mk}}$$

$$b = \sqrt{a}\,.\frac{L}{\sqrt{mk}} \qquad \text{------------------} \ (5)$$

Put the value of equation (5) in equation (4) we get

$$T = \frac{2m\pi a}{L}\left(\sqrt{a}\,.\frac{L}{\sqrt{mk}}\right)$$

$$T = \frac{2\pi m a^{3/2}}{\sqrt{mk}}$$

Apply Squaring on both sides, we get

$$T^2 = \frac{4\pi m^2}{mk}.a^3$$

$$\therefore \qquad T^e \propto a^3$$

This is known as Kepler's 3rd law.

2.11. SCATTERING IN A CENTRAL FORCE FIELD:

RUTHERFORD SCATTERING CROSS – SECTION:

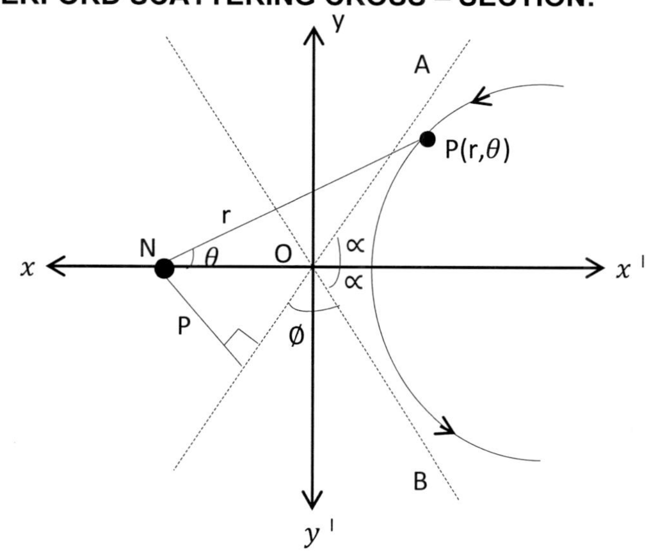

Fig. (2.18)

In Rutherford scattering a positively (+vely) charged particle of charge $z^{\prime}e$ and mass m is scattered by a heavy nucleus N. The nucleus is assumed to at rest during collision. The initial velocity of positively (+vely) charged particle moving towards the heavy nucleus N. As the particle approaches the nucleus the repulsive force is given by

$$F = \frac{q_1 q_2}{4\pi\varepsilon_o r^2}$$

$$F = \frac{(z^{\prime}e)(z\,e)}{4\pi\varepsilon_o r^2}$$

It increases rapidly and the particle changes from a straight line path to a hyperbola having one focus at N, as shown in figure (2.18). The asymptotes AO & BO to the hyperbola give the direction of the incident & scattered particle. Let ∅ be the scattering angle and P be the impact parameter.

$$F = \frac{(z^{\prime}e)(z\,e)}{4\pi\varepsilon_o r^2}$$

$$F = -\frac{1}{r^2}\left[\frac{-(z^{\prime}e)(z\,e)}{4\pi\varepsilon_o}\right]$$

$$F = \frac{-k}{r^2} \qquad \text{------------------ (1)}$$

Where $$k = \frac{-(z^{\prime}e)(z\,e)}{4\pi\varepsilon_o} \qquad \text{----------- (2)}$$

From equation (1) it is clear that the charged particle is moving in a central force field hence equation of path is given by

$$\frac{l}{r} = 1 + e\,Cos\theta \qquad \text{---------------- (3)}$$

Where $$l = \frac{J^2}{mk} \quad \text{------- (4)\&} \qquad e = \sqrt{1 + \frac{2EJ^2}{mk^2}} \quad \text{----- (5)}$$

Now put the value of equation (2) in equation (4), we get

$$l = \frac{J^2}{m\left(\frac{-(z^{\prime}e)(z\,e)}{4\pi\varepsilon_o}\right)}$$

$$l = -\frac{J^2 4\pi\varepsilon_o}{mz^{\prime}z\,e^2} \qquad \text{------------ (6)}$$

Now again put the value of equation (2) in equation (5), we get

$$e = \sqrt{1 + \frac{2EJ^2}{m\left(\frac{-(z^{\prime}e)(z\,e)}{4\pi\varepsilon_o}\right)^2}}$$

$$e = \sqrt{1 + \frac{2EJ^2\,(4\pi\varepsilon_o)^2}{mz^{\prime 2}z^2e^4}} \qquad \text{------------ (7)}$$

Initially the particle is at ∞, so its potential energy (P.E) is given by

$$P.E = \frac{q}{4\pi\varepsilon_o r} = 0$$

Total Energy = Kinetic Energy + Potential Energy

$$E = \frac{1}{2}mV_o^{\ 2} + 0$$

Multiply on both sides with m we get

$$mE = \frac{1}{2}m^2V_o^{\ 2}$$

$$2mE = m^2V_o^{\ 2}$$

$$2mE = (mV_o)^2$$

$$mV_o = \sqrt{2mE} \qquad \text{----------------- (8)}$$

Now $\qquad J = mv\ (\perp^{er}\ \ distance)$

$$J = mV_oP$$

$$J = \sqrt{2mE}.P \qquad \text{---------------- (9)}$$

Now put the value of equation (9) in equation (7), we get

$$e = \sqrt{1 + \frac{2E\left(\sqrt{2mE}.P\right)^2 (4\pi\varepsilon_0)^2}{mz^{\prime 2}z^2e^4}}$$

$$e = \sqrt{1 + \frac{2E\ (2mEP^2)\ (4\pi\varepsilon_0)^2}{mz^{\prime 2}z^2e^4}}$$

$$e = \sqrt{1 + \frac{4E^2P^2\ (4\pi\varepsilon_0)^2}{z^{\prime 2}z^2e^4}} \qquad \text{-------------- (10)}$$

From equation (10) $e > 1$, hence the path of charged particle is hyperbola

Angle of scattering:

The angle between initial and final directions of the positively charged particle is called angle of scattering. i.e., the angle between the asymptotes is called the angle of scattering. In figure (2.18) $\emptyset$ is the angle of scattering.

From Figure (2.18) $\quad \emptyset + \alpha + \alpha = \pi \quad$ (Linear pair)

$$\emptyset + 2\alpha = \pi$$

$$2\alpha = \pi - \emptyset$$

$$\alpha = \frac{\pi}{2} - \frac{\emptyset}{2} \quad \text{-------------- (11)}$$

Using equation (3)

$$\frac{l}{r} = 1 + e\,Cos\theta \quad \text{----------------- (3)}$$

Where

$$r \rightarrow \infty \quad \text{then} \quad \theta = \alpha$$

$$\frac{l}{\infty} = 1 + e\,Cos\alpha$$

$$0 = 1 + e\,Cos\alpha$$

$$e\,Cos\alpha = -1$$

$$Cos\alpha = -\frac{1}{e}$$

$$Cos\left(\frac{\pi}{2} - \frac{\emptyset}{2}\right) = -\frac{1}{e} \quad [From\ eq'n(11)]$$

$$Sin\left(\frac{\emptyset}{2}\right) = -\frac{1}{e}$$

$$Cosec\left(\frac{\emptyset}{2}\right) = -e$$

Squaring on both sides, we get

$$Cosec^2\left(\frac{\emptyset}{2}\right) = e^2$$

$$1 + Cot^2\frac{\emptyset}{2} = \left(\sqrt{1 + \frac{4E^2P^2\,(4\pi\varepsilon_0)^2}{z'^2 z^2 e^4}}\right)^2 \quad [From\ eq'n(10)]$$

$$1 + Cot^2\frac{\emptyset}{2} = 1 + \frac{4E^2P^2\,(4\pi\varepsilon_0)^2}{z'^2 z^2 e^4}$$

$$Cot^2\frac{\emptyset}{2} = \frac{4E^2P^2\,(4\pi\varepsilon_o)^2}{z^{\,\prime 2}z^2e^4}$$

$$Cot\frac{\emptyset}{2} = \frac{2EP\,(4\pi\varepsilon_o)}{z^{\,\prime}z\,e^2} \qquad \text{--------------- (12)}$$

$$P = \frac{z^{\,\prime}z\,e^2}{2E(4\pi\varepsilon_o)}Cot\frac{\emptyset}{2} \qquad \text{--------------- (13)}$$

Now differentiate w.r.to. $\emptyset$, we get

$$\frac{dP}{d\emptyset} = \frac{z^{\,\prime}z\,e^2}{2E(4\pi\varepsilon_o)}\left(-Cosec^2\frac{\emptyset}{2}\right).\frac{1}{2} \qquad \text{------------ (14)}$$

Now Scattering cross – section is given by

$$\sigma(\emptyset) = \frac{-P}{Sin\emptyset}\cdot\frac{dP}{d\emptyset}$$

$$\sigma(\emptyset) = \frac{-\,z^{\,\prime}z\,e^2\,Cot\frac{\emptyset}{2}}{2E(4\pi\varepsilon_o)\left(2\sin\frac{\emptyset}{2}.\ \cos\frac{\emptyset}{2}\right)}\left[\frac{z^{\,\prime}z\,e^2}{2E(4\pi\varepsilon_o)}\left(-Cosec^2\frac{\emptyset}{2}\right).\frac{1}{2}\right]$$

$$\sigma(\emptyset) = \frac{-\,z^{\,\prime}z\,e^2\,\frac{\cos\emptyset/2}{\sin\emptyset/2}}{2E(4\pi\varepsilon_o)\left(2\sin\frac{\emptyset}{2}.\ \cos\frac{\emptyset}{2}\right)}\left[\frac{-\,z^{\,\prime}z\,e^2}{4E(4\pi\varepsilon_o)}\left(Cosec^2\frac{\emptyset}{2}\right)\right]$$

$$\sigma(\emptyset) = \frac{1}{4}\left[\frac{\left(z^{\,\prime}z\,e^2\right)^2}{E^2(4\pi\varepsilon_o)^2 2^2}\right]Cosec^4\frac{\emptyset}{2}$$

$$\sigma(\emptyset) = \frac{1}{4}\left[\frac{z^{\,\prime}z\,e^2}{2E(4\pi\varepsilon_o)}\right]^2 Cosec^4\frac{\emptyset}{2}$$

This is required expression for Rutherford scattering cross – section. This is known as Rutherford's scattering formula for α – particle scattering.

It shows that the scattering cross – section and hence the number of particles scattered must be proportional to

(i) $Cosec^4\frac{\emptyset}{2}$

(ii) the square of the nuclear charge (ze)

(iii) the square of the charge on the particle $(z^{\,\prime}e)$ and

(iv) inversely proportional to the square of the initial kinetic energy i.e., $1/_{E^2}$

Size of Nucleus or distance of closest approach:

$$F = \frac{k\,(z\,'e\,)(z\,e)}{r^2}$$

$$F = \frac{k\,q_1 q_2}{r^2} \qquad \begin{bmatrix} Charge\ on \\ Nucleus = ze = q_1 \\ \alpha - particle = z\,'e = q_2 \end{bmatrix}$$

At distance of closest approach

K. E of α – particle = P.E of α – particle

$$\frac{1}{2}mV^2 = \frac{kze}{r_o}\ X2e$$

$$r_o = \frac{2kze^2}{\frac{1}{2}mV^2}$$

If Z (Gold) = 79 (Atomic number)
E = 1.6 x 10^{-19}c (Charge on proton which is present in nucleus)
m = mα = 6.65 x 10^{-27}kg
V = 2 x 10^7 m/s^{-1}
K = $\frac{1}{9\,X\,10^9}$ Nm^{2}c^{-2}

$$\therefore\ r_o = \frac{2\,X\frac{1}{9\,X\,10^9}X\,79\,X\,\left(1.6\,X\,10^{-19}\right)^2}{\frac{1}{2}X\,665\,X\,10^{-27}X\,(2\,X\,10^{-27})^2}$$

$$r_o = 2.737\,X\,10^{-14}$$

CHAPTER – 3

LAGRANGIAN FORMULATION

In the previous chapters we were able to demonstrate the effectiveness of Newton's laws of motion in solving variety of problems. However, if the system is subject to external constraints, solving the equations of motion may be difficult, and sometimes it may be difficult even to formulate them. The forces of constraints are usually very complex or unknown, which makes the formalism more difficult. To circumvent these difficulties, two different methods, Lagrange's and Hamilton's formulations, have been developed. These techniques use an energy approach and are constructed in such a way that the Newtonian formalism follows from it. Before going over to these procedures, we try to understand certain terms such as constraints, generalized co–ordinates, etc., In this chapter a discussion on the Lagrangian formalism is given.

3.1. CONSTRAINTS

The restrictions or conditions imposed on the motion of particle or system of particles are known as constraints. Hence constraints limit the motion of a particle or system of particles.

Constrained motion:

A motion which can't proceed in any arbitrary manner & obeys certain given conditions is known as constrained motion.

Example 1:

The motion of a rigid body is a constrained motion as distance between any two particles remains constant.
between any two pairs of its particles remains constant.

$$\overrightarrow{OB} + \overrightarrow{BA} = \overrightarrow{OA}$$

$$\vec{r}_j + \vec{C}_{ij} = \vec{r}_i$$

$$\vec{C}_{ij} = \vec{r}_i - \vec{r}_j$$

(Or)

$$\vec{C}_{ij} = \vec{r}_j - \vec{r}_i$$

$$|\vec{C}_{ij}|^2 = |\vec{r}_j - \vec{r}_i|^2$$

$$|\vec{r}_j - \vec{r}_i|^2 - |\vec{C}_{ij}|^2 = 0$$

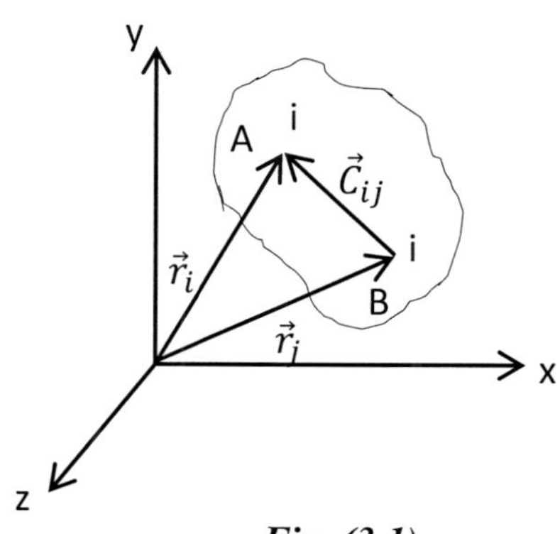

Fig. (3.1)

Example 2:
The motion of a point mass of simple pendulum is constrained because it is subjected to the following two conditions.
a) The point mass oscillates / moves in fixed plane.
b) The point mass always remains at a constant distance from the point of suspension or length of string always remains constant.

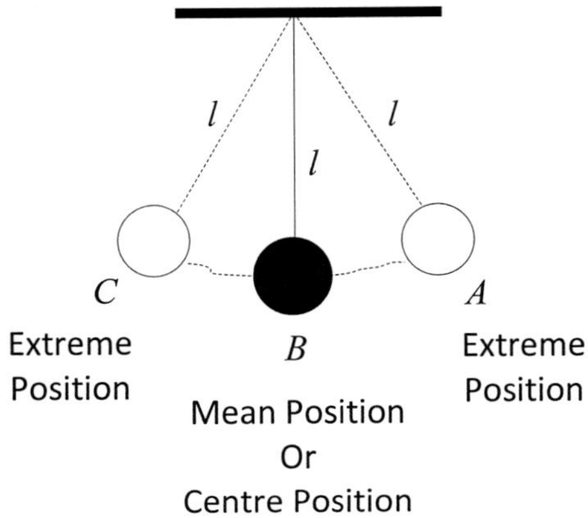

Fig. (3.2)

Example 3:
The motion of a bead sliding down in a wire is simplest example of constrained motion.

Example 4: A particle in a box

Example 5:
A particle placed on the surface of solid sphere is restricted so that it can only move either on the surface or outside the surface.

Types of Constraints
The constraints are classified into 4 types
1) Holonomic constraints
2) Non – Holonomic constraints
3) Scleronomous constraints
4) Rheonomous constraints

1) Holonomic constraints (Integrable constraints):
Definition:
Constraints are said to be Holonomic if constraints imposed on the motion of the system can be expressed in the form of mathematical equation where are the position vectors of system of particles and t is time.

Example 1:

The motion of a rigid body is a constrained motion as distance between any two pairs of its particles remains constant.

$$\overrightarrow{OB} + \overrightarrow{BA} = \overrightarrow{OA}$$

$$\vec{r}_j + \vec{C}_{ij} = \vec{r}_i$$

$$\vec{C}_{ij} = \vec{r}_i - \vec{r}_j$$

(Or)

$$\vec{C}_{ij} = \vec{r}_j - \vec{r}_i$$

$$|\vec{C}_{ij}|^2 = |\vec{r}_j - \vec{r}_i|^2$$

$$|\vec{r}_j - \vec{r}_i|^2 - |\vec{C}_{ij}|^2 = 0$$

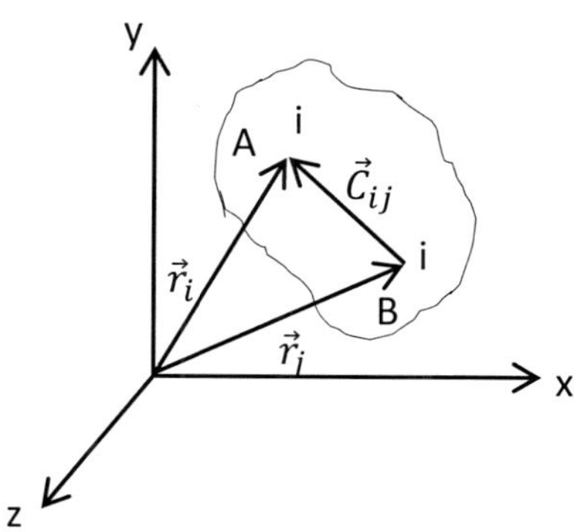

Fig. (3.3)

Example 2:

The sliding of a bead on a circular wire of radius (a) in the xy – plane is another simplest example. The equation of constraint is given by

$$x^2 + y^2 = a^2$$

This can also be expressed in the differential form as

$$xdx + ydy = 0$$

2) Non – Holonomic constraints:

Definition:

Constraints are said to be non – Holonomic if constraint imposed on the motion of a system can't be expressed in the form of a mathematical equation where the position vectors of system of particles are and t is time.

Example 1:

Motion towards $x \le l$
Motion towards $z \le b$
Motion towards $y \le h$

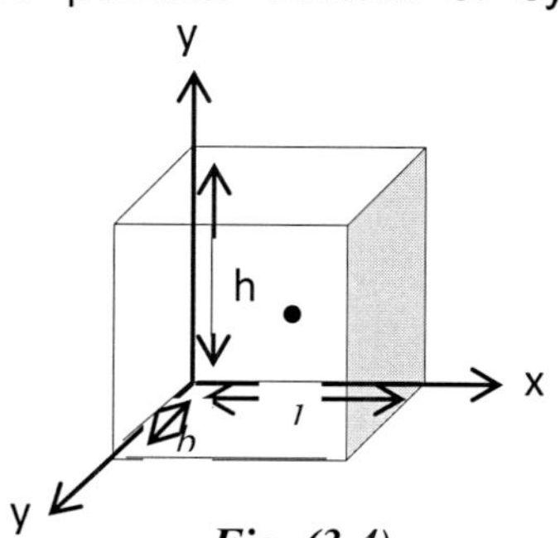

Fig. (3.4)

Example 2:

The motion of particle on the surface of spherical body.

3) Scleronomous constraints:

Definition:

The constraints which are independent of time are known as Scleronomous constraints.

Example 1:

A simple pendulum suspended with a fixed support.

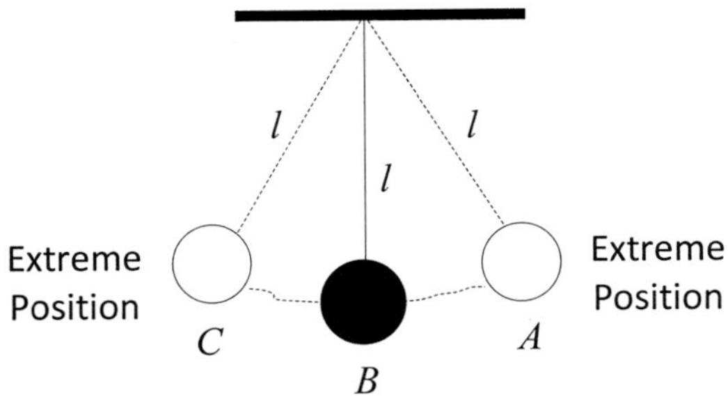

Mean Position Or Centre Position

Fig. (3.5)

4) Rheonomous constraints

Definition:

The constraints which are dependent on time are known as Rheonomous constraints.

Example:

A simple pendulum suspended with a moving support.
i.e., $l = l(t) \rightarrow$ is time dependent

3.2. GENERALIZED CO – ORDINATES

Degrees of Freedom:

The degree of freedom of a mechanical system is that the minimum number of independent co –ordinates required to completely describing its motion.

For Example:

1) When a single free particle moves in space, then 3 position co – ordinates are required to fix its position. So it has 3 degrees of freedom.

Example: A bird flying in sky.
2) When a particle moves in a plane, then it has 2 degrees of freedom
Example: Motion of striker on Carrum Board.

3) When N – Free particle moves in a plane, then it has 3N degrees of freedom.

4) When a system of N – particles subjected to K – constraints, then it has 3N – K degrees of freedom.

$\therefore \qquad f = 3N - K$

5) A bead sliding on a wire.

So, here N = 1
K = 2

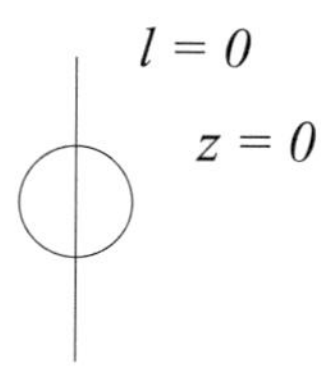

$$\therefore \qquad f = 3N - K$$
$$f = 3(1) - 2$$
$$f = 1$$

∴ Degrees of freedom = 1

6) Motion of simple pendulum

So, here N = 1
K = 2

$$\therefore \qquad f = 3N - K$$
$$f = 3(1) - 2$$
$$f = 1$$

∴ Degrees of freedom = 1

Generalized Co – ordinates:
The degree of freedom of a mechanical system is the minimum no.of independent co – ordinates required to completely describe its motion.

For Example
1) When a single particle moves in space it has 3 degrees of freedom (A bird flying in sky)

2) When a particle moves in a plane, it has 2 degrees of freedom (Motion of striker on Carrum Board)

3) When a particle moves in a straight line, it has 1 degree of freedom (Motion of train on railway track)

Consider a system of N – particles free from constraints has 3N degrees of freedom. If 'k' constraints are imposed on the system then the degree of freedom get reduced to 3N – k = f.

These 'f' no.of minimum independent co – ordinates required to describe configuration & motion of a mechanical system are called generalized co – ordinates and are denoted by $q_1, q_2, ---- q_f$ or $q_i \ (i = 1, 2, ---- f)$ just by.

Transformation Equations

The old co – ordinates can be represented as the function of generalized co – ordinates. If x_i, y_i and z_i are the Cartesian co – ordinates of ith particle of the system, then these Cartesian co – ordinates can be expressed as functions of generalized co – ordinates $q_1, q_2, ---- q_f$ i.e.,

$$\left.\begin{aligned} x_i &= x_i(q_1, q_2, ---- q_f, t) = x_i(q_j, t) \\ y_i &= y_i(q_1, q_2, ---- q_f, t) = y_i(q_j, t) \\ z_i &= z_i(q_1, q_2, ---- q_f, t) = z_i(q_j, t) \end{aligned}\right\} \text{----------- (1)}$$

Where, t denotes time

i = 1, 2, 3, _ _ _ _ N

j = 1, 2, 3, _ _ _ _ n

If r_i is the position vector of ith particle i.e., $r_i = ix_i + jy_i + kz_i$ where i, j, k are unit vectors along x, y and z axes respectively, then

$$\vec{r}_i = \vec{r}_i(q_1, q_2, ---- q_f, t) \qquad \text{------------- (2)}$$

This is the vector form of equation (1)

The equations like (1) & (2) are called ***transformation equations***. The functions and their derivatives in the above two equations are supposed to be continues. The equations also contain the constraints explicitly.

The variable q_j may not necessarily have the dimension of length. These may be angle such as θ & $\emptyset$, in case of spherical polar co – ordinates.

The inverse transformation equations of equation (1) can be written as

$$q_j = q_j\,[x_1, y_1, z_1\ , x_2, y_2, z_2 --- x_N, y_N, z_N, t]$$
$$q_j = q_j\,[r_1, r_2, r_3 ---- r_N, t]$$
$$q_j = q_j\,[r_i, t] \quad \text{----------------- (3)}$$

Where i = 1, 2, 3, _ _ _ _ N

j = 1, 2, 3, _ _ _ _ n

The time derivative of generalized co – ordinates namely $\dot{q}_j = \frac{dq_j}{dt}$ are generalized velocities.

Example (1):

If a particle of mass (m) is constrained to move on a circular wire in a plane, then motion can be described by generalized co – ordinate (θ)

$$x = rcos\theta\, y = rsin\theta$$

$$\frac{rsin\theta}{rcos\theta} = \frac{y}{x}$$

$$tan\theta = \frac{y}{x}$$

$$\theta = \tan^{-1}\left(\frac{y}{x}\right)$$

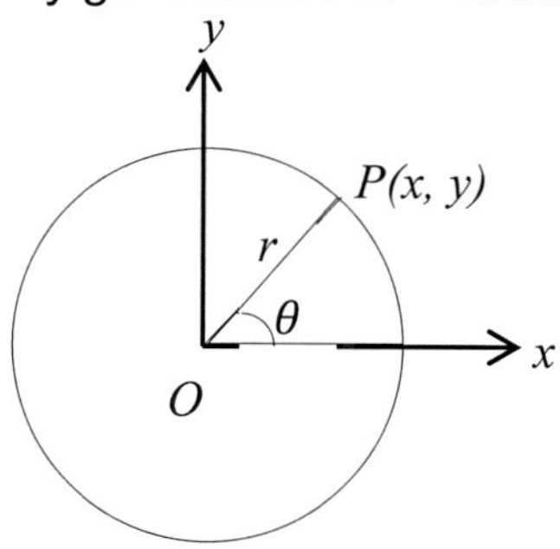

Fig. (3.6)

Example (2):

Motion of simple pendulum oscillating in vertical plane can be described in terms of Cartesian co – ordinates x, y, z

$$x = l\,cos\theta \quad y = l\,sin\theta$$

Where

l – is constant

θ – is variable (Generalized co – ordinates)

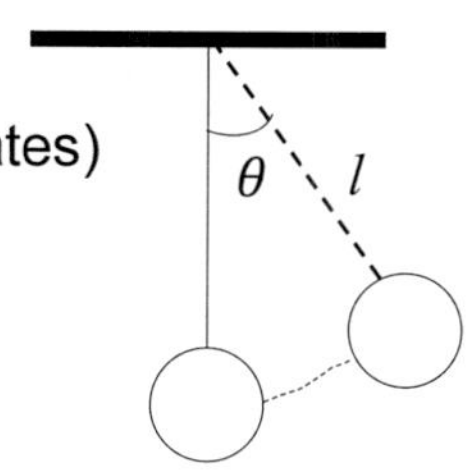

Fig. (3.7)

Thus,

$$q_1 = \theta = \cos^{-1}\left(\frac{x}{l}\right)$$

$$q_2 = \theta = \sin^{-1}\left(\frac{y}{l}\right)$$

Configuration Space:

For a free particle we need 3 co – ordinates to explain it's configuration. i.e., the space for the description of motion of a single particle must be three dimensional.

For two free particles we need 6 co – ordinates to explain it's configuration. i.e., for the description of motion of two particles, we requires six – dimensional space.

For N particles we need 3N co – ordinates to explain it's configuration. Thus 3N independent co – ordinates or degrees of freedom in total. Equivalently we can describe the case of N free particle as a single particle moving in 3N dimensional space.

If the system is such that the conditions of constraints are expressed by k equations, then there will be 3N – k = f independent co – ordinates or degree of freedom. In this case the problem of N particles can be considered as that of a single particle in f dimensional space.

This 3N dimensional spaces in the absence of constraints and f = 3N – k dimensional space in the presence of constraints in the case of system of particles is called ***configuration space***, where the conditions of constraints are expressed by k equations.

3.3. GENERALIZED NOTATIONS

Now we shall express displacement, velocity, acceleration, momentum, force and potential energy in terms of generalized co – ordinates.

1) Generalized Displacement:

Let us consider a N – particle system for which a small displacement $\delta\vec{r}_i$ is defined by change in position co – ordinates $\vec{r}_i$ $(i = 1, 2, 3, - - -N)$ with time (t) kept as constant. The position vector $\vec{r}_i$ of the ith particle in the form of generalized co – ordinates can be written as

$$\vec{r}_i = \vec{r}_i(q_1, q_2, - - - - q_f, t)$$

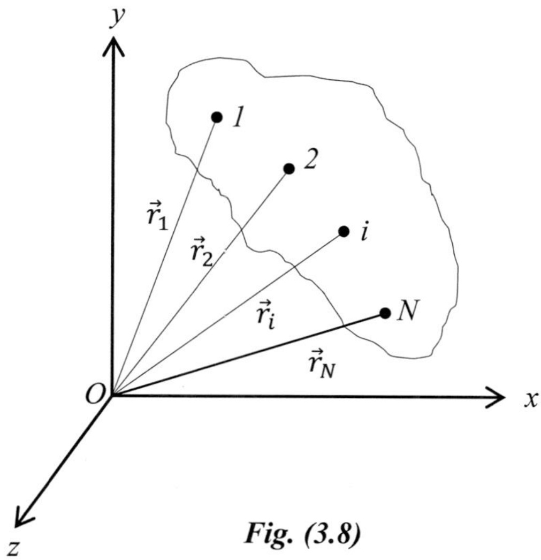

Fig. (3.8)

$$\left[\begin{array}{c} We\ know\ that \\ f(\vec{r}) = f(x, y, z) \\ df = \frac{\partial f}{\partial x}dx + \frac{\partial f}{\partial y}dy + \frac{\partial f}{\partial z}dz \\ \delta f = \frac{\partial f}{\partial x}\delta x + \frac{\partial f}{\partial y}\delta y + \frac{\partial f}{\partial z}\delta z \\ This\ is\ Euler's\ Theorem \end{array}\right]$$

Using Euler's Theorem, we have

$$\delta\vec{r}_i = \frac{\partial\vec{r}_i}{\partial q_1}\delta q_1 + \frac{\partial\vec{r}_i}{\partial q_2}\delta q_2 + - - - - - - + \frac{\partial\vec{r}_i}{\partial q_f}\delta q_f$$

$$\delta\vec{r}_i = \sum_{k=1}^{f}\frac{\partial\vec{r}_i}{\partial q_k}\delta q_k$$

Where δq_k represents Generalized Displacement.

2) Generalized Velocity:

Let us consider a N – particle system, the generalized co – ordinate $\dot{q}_k$ is the time derivative of the generalized co – ordinate q_k. The position vector $\vec{r}_i$ of the ith particle in the form of generalized co – ordinate & time (t) can be written as

$$\vec{r}_i = \vec{r}_i(q_1, q_2, - - - - q_f, t)$$

Using Euler's Theorem, we have

$$d\vec{r}_i = \frac{\partial\vec{r}_i}{\partial q_1}dq_1 + \frac{\partial\vec{r}_i}{\partial q_2}dq_2 + - - - - - - + \frac{\partial\vec{r}_i}{\partial q_f}\delta q_f + \frac{\partial\vec{r}_i}{\partial t}dt$$

Dividing on both sides by dt, we have

$$\frac{d\vec{r}_i}{dt} = \frac{\partial \vec{r}_i}{\partial q_1}\frac{dq_1}{dt} + \frac{\partial \vec{r}_i}{\partial q_2}\frac{dq_2}{dt} + ------ + \frac{\partial \vec{r}_i}{\partial q_f}\frac{\delta q_f}{dt} + \frac{\partial \vec{r}_i}{\partial t}\frac{dt}{dt}$$

$$\vec{V}_i = \sum_{k=1}^{f}\frac{\partial \vec{r}_i}{\partial q_k}\frac{\partial q_k}{dt} + \frac{\partial \vec{r}_i}{\partial t}$$

$$\vec{V}_i = \sum_{k=1}^{f}\frac{\partial \vec{r}_i}{\partial q_k}\dot{q}_k + \frac{\partial \vec{r}_i}{\partial t}$$

Where $\dot{q}_k$ is called generalized Velocity.

3) Generalized Acceleration:

In generalized acceleration is equal to the time derivative of the generalized velocity. Let us consider a system of N – particles, the velocity of ith particle of the system is given by

$$\vec{V}_i = \dot{\vec{r}}_i = \sum_{k=1}^{f}\frac{\partial \vec{r}_i}{\partial q_k}\dot{q}_k + \frac{\partial \vec{r}_i}{\partial t} \qquad --------------- (1)$$

Differentiating on both sides w. r. to time, we get

$$\dot{\vec{V}}_i = \ddot{\vec{r}}_i = \frac{d}{dt}\left[\sum_{k=1}^{f}\frac{\partial \vec{r}_i}{\partial q_k}\dot{q}_k + \frac{\partial \vec{r}_i}{\partial t}\right]$$

$$\vec{a}_i = \frac{d}{dt}\sum_{k=1}^{f}\frac{\partial \vec{r}_i}{\partial q_k}\dot{q}_k + \frac{d}{dt}\frac{\partial \vec{r}_i}{\partial t}$$

$$\left[\begin{matrix} We\ know\ that \\ \frac{d\,(uv)}{dx} = u\frac{dv}{x} + v\frac{du}{dx}\end{matrix}\right]$$

$$\vec{a}_i = \sum_{k=1}^{f}\frac{\partial \dot{\vec{r}}_i}{\partial q_k}\dot{q}_k + \sum_{k=1}^{f}\frac{\partial \vec{r}_i}{\partial q_k}\ddot{q}_k + \frac{\partial \dot{\vec{r}}_i}{\partial t} \qquad ----------------- (2)$$

Put the value of equation (1) in equation (2) we get

$$\vec{a}_i = \sum_{k=1}^{f}\frac{\partial}{\partial q_k}\left[\sum_{k=1}^{f}\frac{\partial \vec{r}_i}{\partial q_k}\dot{q}_k + \frac{\partial \vec{r}_i}{\partial t}\right]\dot{q}_k + \sum_{k=1}^{f}\frac{\partial \vec{r}_i}{\partial q_k}\ddot{q}_k + \frac{\partial}{\partial t}\left[\sum_{k=1}^{f}\frac{\partial \vec{r}_i}{\partial q_k}\dot{q}_k + \frac{\partial \vec{r}_i}{\partial t}\right]$$

$$\vec{a}_i = \sum_{k=1}^{f}\sum_{j=1}^{f}\frac{\partial^2 \vec{r}_i}{\partial \dot{q}_k \partial q_j}\dot{q}_j\dot{q}_k + \sum_{k=1}^{f}\frac{\partial^2 \vec{r}_i}{\partial q_k \partial t}\dot{q}_k + \sum_{k=1}^{f}\frac{\partial \vec{r}_i}{\partial q_k}\ddot{q}_k + \sum_{j=1}^{f}\frac{\partial^2 \vec{r}_i}{\partial t\,\partial q_j} + \frac{\partial^2 \vec{r}_i}{\partial t^2}$$

$$----------------- (3)$$

This is the required expression for Generalized Acceleration. Here $\ddot{q}_k$ is the Generalized Acceleration.

Equation (4) shows that Cartesian components of acceleration are not the linear functions of generalized acceleration $\ddot{q}_k$ alone, but depend quadratically and linearly on generalized velocity components $\dot{q}_k$ as well. However, according to Lagrange's approach the computation of second derivatives of generalized co – ordinates is not essential.

4) Generalized Momentum:

The kinetic energy of the ith particle having mass m_i & velocity V_i is given by

$$T = \frac{1}{2}m_i {v_i}^2$$

$$T = \frac{1}{2}m_i \left(\frac{dx_i}{dt}\right)^2 \qquad \left[V_i = \frac{dx_i}{dt} = \dot{x}_i\right]$$

$$T = \frac{1}{2}m_i(\dot{x}_i)^2 \qquad \text{-------------- (1)}$$

Now, partially differentiating equation (1) w.r.to $\dot{x}_i$, we get

$$\frac{\partial T}{\partial \dot{x}_i} = \frac{1}{2}\, m_i 2\dot{x}_i$$

$$\frac{\partial T}{\partial \dot{x}_i} = m_i\dot{x}_i \qquad \text{--------------- (2)}$$

Now linear momentum P_i of ith particle is given by

$$P_i = m_i v_i$$

$$P_i = m_i\, \frac{dx_i}{dt}$$

$$P_i = m_i\dot{x}_i \qquad \text{----------------- (3)}$$

Comparing equation (2) & (3) , we get

$$P_i = \frac{\partial T}{\partial \dot{x}_i} = m_i\dot{x}_i \qquad \text{------------------ (4)}$$

Similarly linear momentum associated with generalized co – ordinates q_k called generalized momentum P_k is given by

$$P_k = \frac{\partial T}{\partial \dot{q}_k} = m_k\dot{q}_k \qquad \text{------------------ (5)}$$

5) Generalized Force:

Consider a system having N – particles & no constraints are there on the system. Suppose a force $\sum_{i=1}^{N} \vec{F}_i$ is acting on the system causing an arbitrary small displacement $\delta\vec{r}_i$ of the system by doing work δw is given by

$$\delta w = \sum_{i=1}^{N} \vec{F}_i \, . \delta\vec{r}_i \qquad \text{-------------- (1)}$$

We know that

$$\delta\vec{r}_i = \sum_{i=1}^{3N} \frac{\partial \vec{r}_i}{\partial q_j} \delta q_j \qquad \text{-------- (2)} \qquad \left[\begin{array}{c} \text{Here no constraints are applied} \\ i.e., k = 0 \\ \therefore f = 3N - k \\ f = 3N - 0 \\ f = 3N \\ \text{so, we are replacing } 3N \text{ in place of } f \end{array}\right]$$

Put the value of equation (2) in equation (1), we get

$$\delta w = \sum_{i=1}^{N} \vec{F}_i \, . \sum_{i=1}^{3N} \frac{\partial \vec{r}_i}{\partial q_j} \delta q_j$$

$$\delta w = \sum_{j=1}^{3N} Q_j . \, \delta q_j \qquad \left[\begin{array}{c} \text{Here} \\ Q_j = \sum_{i=1}^{N} \vec{F}_i \, . \frac{\partial \vec{r}_i}{\partial q_j} \end{array}\right]$$

Here Q_j is called Generalized Force, associated with co – ordinate q_k. It may be noted that the expressions for work done in Cartesian co – ordinates W = $\sum_k \vec{F}_i \, . \, \delta r_i$ and in generalized co – ordinates $\sum_k Q_k \delta q_k$ are analogous. It may be noted that it is not necessary that the dimensions of $Q_k$are that of force; but it is certain that the product of generalized force and generalized displacement has the dimensions of work.

Virtual work:

Work done by force on a particle due to it's virtual displacement is called virtual work.

3.4. PRINCIPLE OF VIRTUAL WORK

Statement:

If a system is in equilibrium then the virtual work for system due to applied force will vanish (or) zero.

Suppose a system is in equilibrium i.e., total force on each particle is zero. i.., $\quad \vec{F}_i = 0$, Then clearly $\quad \vec{F}_i \delta\vec{r}_i = 0$

Where, $\quad \delta\vec{r}_i$ – is virtual displacement

The sum over all the particles must be zero.

i.e., $\sum_i \vec{F_i}\,\delta\vec{r_i} = 0$

Here $\vec{F_i} = \vec{F_i}^{\,a} + \vec{f_i}$

Where

$\vec{F_i}^{\,a}$ - is applied force

$\vec{f_i}$ – is constraints force

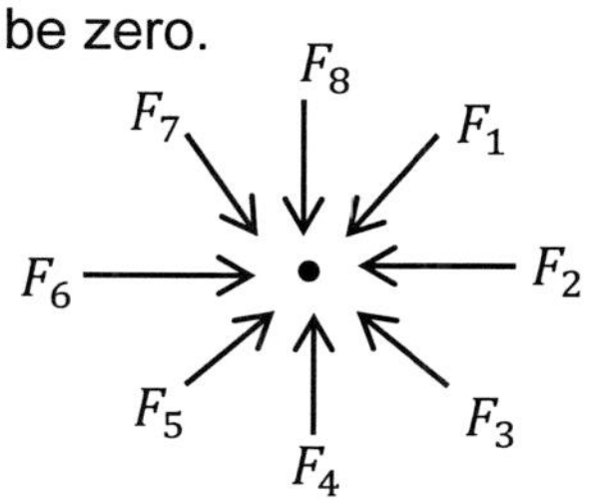

Fig. (3.9)

$$\sum_i \vec{F_i}^{\,a}\,\delta\vec{r_i} + \sum_i \vec{f_i}\,\delta\vec{r_i} = 0$$

Assume that we consider only those systems in which the net work done by the forces of constraints is zero.

$$\sum_i \vec{F_i}^{\,a}\,\delta\vec{r_i} = 0$$

i.e., $\sum_i \vec{f_i}\,\delta\vec{r_i} = 0$

3.5. D'ALEMBERT'S PRINCIPLE

The principle of virtual work stated that it is applicable to the statics of a system of particles, but it can be related to obtain a similar principle of dynamics.

According to Newton's second law of motion, the force is defined as rate of change of momentum. i.e.,

$$F_i = \frac{dP_i}{dt}$$

$$F_i = d\dot{P_i}$$

Where $\dot{P_i}$ – is momentum of ith particle

F_i – is force acting on i[th] particle

∴ $F_i - \dot{P_i} = 0$ ------------------- (1)

This states that the particles in the system will be in equilibrium with a force equal to the actual force plus a reverse effective force $-\vec{\dot{P_i}}$

Now multiplying on both sides of equation (1) with $\delta\vec{r_i}$ we get

$$\left(F_i - \vec{\dot{P_i}}\right)\delta\vec{r_i} = 0$$

Foe the whole system

$$\sum_i \left(F_i - \dot{\vec{P}_i}\right) \delta\vec{r}_i = 0 \qquad \text{---------} (2)$$

If the forces of constraints $\vec{f}_i$ are present then we have

$$\vec{F}_i = \vec{F}_i^{\,a} + \vec{f}_i \qquad \text{-----------------} (3)$$

Put the value of $\vec{F}_i$ in equation (2) we get

$$\sum_i \left[\left(\vec{F}_i^{\,a} + \vec{f}_i\right) - \dot{\vec{P}_i}\right] \delta\vec{r}_i = 0$$

$$\sum_i \left[\left(\vec{F}_i^{\,a} - \dot{\vec{P}_i}\right) + \vec{f}_i\right] \delta\vec{r}_i = 0$$

$$\sum_i \left(\vec{F}_i^{\,a} - \dot{\vec{P}_i}\right) \delta\vec{r}_i + \sum_i \vec{f}_i \,.\, \delta\vec{r}_i = 0 \qquad \text{---------------} (4)$$

Consider only those systems for which the net virtual work done by the constraint forces is zero.

i.e., $\quad \sum_i \vec{f}_i \,.\, \delta\vec{r}_i = 0$

Now, Equation (4) reduces to

$$\sum_i \left(\vec{F}_i^{\,a} - \dot{\vec{P}_i}\right) \delta\vec{r}_i = 0 \qquad \text{--------------} (5)$$

This equation (5) is called D'Alembert's principle.

3.6. LAGRANGE'S EQUATIONS FROM D'ALEMBERT'S PRINCIPLE

Lagrange's equation of motion for a Holonomic system by using D'Alembert's principle:

Consider a system of N – particles, according to transformation equations the position vectors of the particles $r_1, r_2, r_3 --- r_i ---$ are expressed as the functions of generalized co – ordinates $q_1, q_2, q_3 --- q_k --- q_f$ and the time (t). i.e.,

$$r_1 = r_1(q_1, q_2, q_3 --- q_k --- q_f, t)$$
$$r_2 = r_2(q_1, q_2, q_3 --- q_k --- q_f, t)$$

....

....

....

$$r_i = r_i(q_1, q_2, q_3 --- q_k --- q_f, t) \qquad \text{-------------} (1)$$

The virtual displacement $\delta\vec{r}_i$ is given by

$$\delta\vec{r}_i = \frac{\partial\vec{r}_i}{\partial q_1}\delta q_1 + \frac{\partial\vec{r}_i}{\partial q_2}\delta q_2 + --- + \frac{\partial\vec{r}_i}{\partial q_k}\delta q_k + --- + \frac{\partial\vec{r}_i}{\partial q_f}\delta q_f$$

$$\delta\vec{r}_i = \sum_{k=1}^{f}\frac{\partial\vec{r}_i}{\partial q_k}\delta q_k \qquad \text{------------ (2)}$$

According to D'Alembert's principle, we have

$$\sum_{i=1}^{N}\left(\vec{F}_i - \dot{\vec{P}}_i\right)\delta\vec{r}_i = 0$$

$$\sum_{i=1}^{N}\vec{F}_i.\delta\vec{r}_i - \sum_{i=1}^{N}\dot{\vec{P}}_i.\delta\vec{r}_i = 0$$

$$\sum_{i=1}^{N}\vec{F}_i.\delta\vec{r}_i = \sum_{i=1}^{N}\dot{\vec{P}}_i.\delta\vec{r}_i \qquad \text{-------------- (3)}$$

Let us first take LHS of equation (3)

$$\sum_{i=1}^{N}\vec{F}_i.\delta\vec{r}_i = \sum_{i=1}^{N}\vec{F}_i.\left[\sum_{k=1}^{f}\frac{\partial\vec{r}_i}{\partial q_k}\delta q_k\right]$$

$$\sum_{i=1}^{N}\vec{F}_i.\delta\vec{r}_i = \sum_{i=1}^{f}\left[\sum_{i=1}^{N}\vec{F}_i\frac{\partial\vec{r}_i}{\partial q_k}\right]\delta q_k$$

$$\sum_{i=1}^{N}\vec{F}_i.\delta\vec{r}_i = \sum_{i=1}^{f}Q_k\delta q_k \qquad \text{------------ (4)}$$

Where, $$Q_k = \sum_{i=1}^{N}\vec{F}_i\frac{\partial\vec{r}_i}{\partial q_k} \qquad \text{--------------- (5)}$$

This is the generalized force.

If the system is conservative then force $\vec{F}_i$ is derivable from scalar potential V.

i.e., $$\vec{F}_i = -\nabla_i V \qquad \left[\begin{matrix}We\ know\ that\\ \vec{r}_i = x_i\hat{\imath} + y_i\hat{\jmath} + z_i\hat{k}\end{matrix}\right]$$

$$\vec{F}_i = -\left(\hat{\imath}\frac{\partial}{\partial x_i} + \hat{\jmath}\frac{\partial}{\partial y_i} + \hat{k}\frac{\partial}{\partial z_i}\right)V$$

$$\vec{F}_i = -\left(\hat{\imath}\frac{\partial V}{\partial x_i} + \hat{\jmath}\frac{\partial V}{\partial y_i} + \hat{k}\frac{\partial V}{\partial z_i}\right) \qquad \text{------------- (6)}$$

Put the value of equation (6) in equation (5) we get

$$Q_k = \sum_{i=1}^{N} - \left(\hat{\imath}\frac{\partial V}{\partial x_i} + \hat{\jmath}\frac{\partial V}{\partial y_i} + \hat{k}\frac{\partial V}{\partial z_i}\right) . \frac{\partial \vec{r}_1}{\partial q_k}$$

$$Q_k = \sum_{i=1}^{N} - \left(\hat{\imath}\frac{\partial V}{\partial x_i} + \hat{\jmath}\frac{\partial V}{\partial y_i} + \hat{k}\frac{\partial V}{\partial z_i}\right) . \frac{\partial}{\partial q_k}\left(x_i\hat{\imath} + y_i\hat{\jmath} + z_i\hat{k}\right)$$

$$Q_k = -\sum_{i=1}^{N} \left(\hat{\imath}\frac{\partial V}{\partial x_i} + \hat{\jmath}\frac{\partial V}{\partial y_i} + \hat{k}\frac{\partial V}{\partial z_i}\right) . \left(\hat{\imath}\frac{\partial x_i}{\partial q_k} + \hat{\jmath}\frac{\partial y_i}{\partial q_k} + \hat{k}\frac{\partial z_i}{\partial q_k}\right)$$

$$Q_k = -\sum_{i=1}^{N} \left(\frac{\partial V}{\partial x_i} . \frac{\partial x_i}{\partial q_k} + \frac{\partial V}{\partial y_i} . \frac{\partial y_i}{\partial q_k} + \frac{\partial V}{\partial z_i} . \frac{\partial z_i}{\partial q_k}\right)$$

$$Q_k = -\frac{\partial V}{\partial q_k} \qquad \text{------------------ (7)}$$

Put the value of equation (7) in equation (4) we get

$$\sum_{i=1}^{N} \vec{F}_i . \delta\vec{r}_i = -\sum_{i=1}^{f} \frac{\partial V}{\partial q_k}\delta q_k \qquad \text{--------------- (8)}$$

Let us now take RHS of equation (3) i.e.,

$$\sum_{i=1}^{N} \dot{\vec{P}}_i . \delta\vec{r}_i = \sum_{i=1}^{N} m_i \dot{\vec{V}}_i \left[\sum_{k=1}^{f} \frac{\partial \vec{r}_i}{\partial q_k}\delta q_k\right]$$

$$\sum_{i=1}^{N} \dot{\vec{P}}_i . \delta\vec{r}_i = \sum_{i=1}^{N} m_i \ddot{\vec{r}}_i \left[\sum_{k=1}^{f} \frac{\partial \vec{r}_i}{\partial q_k}\delta q_k\right]$$

$$\sum_{i=1}^{N} \dot{\vec{P}}_i . \delta\vec{r}_i = \sum_{i=1}^{f} \left[\sum_{i=1}^{N} m_i \ddot{\vec{r}}_i \frac{\partial \vec{r}_i}{\partial q_k}\right]\delta q_k \qquad \text{--------------- (9)}$$

Let us now consider

$$\frac{d}{dt}\left(m_i\dot{\vec{r}}_i\frac{\partial \vec{r}_i}{\partial q_k}\right) = m_i\ddot{\vec{r}}_i\frac{\partial \vec{r}_i}{\partial q_k} + m_i\dot{\vec{r}}_i\,\frac{d}{dt}\left(\frac{\partial \vec{r}_i}{\partial q_k}\right)$$

$$\frac{d}{dt}\left(m_i\dot{\vec{r}}_i\frac{\partial \vec{r}_i}{\partial q_k}\right) - m_i\dot{\vec{r}}_i\,\frac{d}{dt}\left(\frac{\partial \vec{r}_i}{\partial q_k}\right) = m_i\ddot{\vec{r}}_i\frac{\partial \vec{r}_i}{\partial q_k} \qquad \text{-------------- (10)}$$

Put the value of equation (10) in equation (9) we get

$$\sum_{i=1}^{N} \dot{\vec{P}}_i . \delta\vec{r}_i = \sum_{i=1}^{f} \left\{\sum_{i=1}^{N}\left[\frac{d}{dt}\left(m_i\dot{\vec{r}}_i\frac{\partial \vec{r}_i}{\partial q_k}\right) - m_i\dot{\vec{r}}_i\,\frac{d}{dt}\left(\frac{\partial \vec{r}_i}{\partial q_k}\right)\right]\right\}\delta q_k$$

$$\sum_{i=1}^{N} \dot{\vec{P}}_i . \delta\vec{r}_i = \sum_{i=1}^{f} \left\{\sum_{i=1}^{N}\left[\frac{d}{dt}\left(m_i\dot{\vec{r}}_i\frac{\partial \vec{r}_i}{\partial q_k}\right) - m_i\dot{\vec{r}}_i\,\frac{\partial}{\partial q_k}\left(\frac{d\vec{r}_i}{dt}\right)\right]\right\}\delta q_k \qquad \text{----- (11)}$$

$$\left.\begin{aligned} &\text{Now } \vec{V}_i = \frac{d\vec{r}_i}{dt} \quad \text{Also } \frac{d\vec{r}_i}{dq_k} = \frac{d\dot{\vec{r}}_i}{d\dot{q}_k} \\ &\vec{V}_i = \dot{\vec{r}}_i \qquad\qquad \frac{d\vec{r}_i}{dq_k} = \frac{d\vec{V}_i}{d\dot{q}_k} \end{aligned}\right\} \qquad \text{---------------- (12)}$$

Put the value of equation (12) in equation (11) we get

$$\sum_{i=1}^{N} \dot{\vec{P}}_i.\delta\vec{r}_i = \sum_{i=1}^{f}\left\{\sum_{i=1}^{N}\left[\frac{d}{dt}\left(m_i\vec{V}_i\frac{\partial\vec{V}_i}{\partial\dot{q}_k}\right) - m_i\vec{V}_i\frac{\partial\vec{V}_i}{\partial q_k}\right]\right\}\delta q_k$$

$$\sum_{i=1}^{N} \dot{\vec{P}}_i.\delta\vec{r}_i = \sum_{i=1}^{f}\left\{\sum_{i=1}^{N}\frac{d}{dt}\left(m_i\vec{V}_i\frac{\partial\vec{V}_i}{\partial\dot{q}_k}\right) - \sum_{i=1}^{N} m_i\vec{V}_i\frac{\partial\vec{V}_i}{\partial q_k}\right\}\delta q_k$$

$$\sum_{i=1}^{N} \dot{\vec{P}}_i.\delta\vec{r}_i = \sum_{i=1}^{f}\left\{\frac{d}{dt}\sum_{i=1}^{N}\left(m_i\vec{V}_i\frac{\partial\vec{V}_i}{\partial\dot{q}_k}\right) - \sum_{i=1}^{N} m_i\vec{V}_i\frac{\partial\vec{V}_i}{\partial q_k}\right\}\delta q_k$$

$$\left[\begin{aligned} &\frac{\partial}{\partial\dot{q}_k}\left(\sum_{i=1}^{N}\frac{1}{2}m_i V_i^2\right) \\ &= \sum_{i=1}^{N}\frac{1}{2}m_i\, 2V_i\frac{\partial V_i}{\partial\dot{q}_k} \\ &= \sum_{i=1}^{N} m_i V_i\frac{\partial V_i}{\partial\dot{q}_k} \end{aligned}\right]$$

$$\sum_{i=1}^{N} \dot{\vec{P}}_i.\delta\vec{r}_i = \sum_{i=1}^{f}\left\{\frac{d}{dt}\left[\frac{\partial}{\partial\dot{q}_k}\left(\sum_{i=1}^{N}\frac{1}{2}m_i V_i^2\right)\right] - \frac{\partial}{\partial q_k}\sum_{i=1}^{N}\frac{1}{2}m_iV_i^2\right\}\delta q_k$$

$$\sum_{i=1}^{N} \dot{\vec{P}}_i.\delta\vec{r}_i = \sum_{i=1}^{f}\left\{\frac{d}{dt}\left[\frac{\partial}{\partial\dot{q}_k}T\right] - \frac{\partial}{\partial q_k}T\right\}\delta q_k \qquad \text{--------------- (13)}$$

Where $\quad T = \sum_{i=1}^{N}\frac{1}{2}m_i V_i^2$ K.E of whole system

Put the values of equation (8) & (13) in equation (3) we get

$$-\sum_{i=1}^{f}\frac{\partial V}{\partial q_k}\delta q_k = \sum_{i=1}^{f}\left\{\frac{d}{dt}\left[\frac{\partial}{\partial\dot{q}_k}T\right] - \frac{\partial}{\partial q_k}T\right\}\delta q_k$$

$$-\sum_{i=1}^{f}\frac{\partial V}{\partial q_k}\delta q_k - \sum_{i=1}^{f}\left\{\frac{d}{dt}\left[\frac{\partial}{\partial\dot{q}_k}T\right] - \frac{\partial}{\partial q_k}T\right\}\delta q_k = 0$$

$$\sum_{i=1}^{f}\frac{\partial V}{\partial q_k}\delta q_k + \sum_{i=1}^{f}\left\{\frac{d}{dt}\left[\frac{\partial}{\partial\dot{q}_k}T\right] - \frac{\partial}{\partial q_k}T\right\}\delta q_k = 0$$

$$\sum_{i=1}^{f}\left[\frac{\partial V}{\partial q_k} + \frac{d}{dt}\left(\frac{\partial}{\partial\dot{q}_k}T\right) - \frac{\partial}{\partial q_k}T\right]\delta q_k = 0$$

$$\frac{\partial V}{\partial q_k} + \frac{d}{dt}\left(\frac{\partial}{\partial\dot{q}_k}T\right) - \frac{\partial}{\partial q_k}T = 0$$

$$\frac{d}{dt}\left(\frac{\partial}{\partial \dot{q}_k}T\right) - \frac{\partial}{\partial q_k}T + \frac{\partial V}{\partial q_k} = 0$$

$$\frac{d}{dt}\left(\frac{\partial}{\partial \dot{q}_k}T\right) - \frac{\partial}{\partial q_k}(T - V) = 0$$

$$\frac{d}{dt}\left[\frac{\partial}{\partial \dot{q}_k}(T - V)\right] - \frac{\partial}{\partial q_k}(T - V) = 0 \qquad \left[\frac{\partial}{\partial \dot{q}_k}(T - V) = \frac{\partial}{\partial \dot{q}_k}T\right]$$

Where $L = T - V$ Lagrangian of the system

$$\boldsymbol{\frac{d}{dt}\left(\frac{\partial L}{\partial \dot{q}_k}\right) - \frac{\partial L}{\partial q_k} = 0}$$

This is the required Lagrange's equation of motion for Holonomic system.

3.7. LAGRANGE'S EQUATIONS FOR SYSTEMS CONTAINING DISSIPATIVE FORCES

Consider any system of particles containing dissipative (or frictional) force. It is the experimental fact that in general the dissipative (or frictional) force is proportional to the velocity of the particles. i.,

$$F_i^{\,d} = -\lambda_i \dot{r}_i \qquad \text{------------------ (1)}$$

Where,
$F_i^{\,d}$ - Denotes by the dissipative force
$\dot{r}_i$ – is the velocity of ith particle
λ_i – is the corresponding constant of proportionality.

The dissipative (or frictional) forces, which are proportional to velocity, ay be derived in terms of a new quantity called the *Rayleigh's dissipation function*, which is defined as

$$R = \frac{1}{2}\sum_i \lambda_i \dot{r}_i^{\,2} \qquad \text{------------------ (2)}$$

Where the summation I taken over all the particles of the system, which gives

$$\frac{\partial R}{\partial \dot{r}_i} = \lambda_i \dot{r}_i = -F_i^{\,d} \qquad (From\ equation\ 1)$$

Or $$F_i^{\,d} = -\frac{\partial R}{\partial \dot{r}_i} \qquad \text{------------------ (3)}$$

The Lagrange's equation in terms of r_i is given by

$$\frac{d}{dt}\left(\frac{\partial L}{\partial r_i}\right) - \frac{\partial L}{\partial r_i} + \frac{\partial R}{\partial r_i} = 0 \qquad \text{------------------ (4)}$$

Where term $\frac{\partial R}{\partial r_i}$ represents the dissipative (or frictional) force.
In order to find the Lagrange's equation, in generalized co – ordinates, we have to find the components of generalized force resulting from the dissipative (or frictional) force.

If ${Q_k}^d$ is the component of generalized force, then

$${Q_k}^d = \sum_i {F_i}^d . \frac{\partial r_i}{\partial q_k}$$

$${Q_k}^d = \sum_i -\lambda_i \dot{r}_i \cdot \frac{\partial r_i}{\partial q_k} \qquad (From\ equation\ 1)$$

$${Q_k}^d = -\sum_i \lambda_i \dot{r}_i \cdot \frac{\partial \dot{r}_i}{\partial \dot{q}_k}$$

$$\left[Since\ \frac{\partial r_i}{\partial q_k} = \frac{\partial \dot{r}_i}{\partial \dot{q}_k}\ from\ equation\ (12)\ of\ section\ 3.6\right]$$

$${Q_k}^d = -\sum_i \lambda_i \frac{\partial}{\partial \dot{q}_k}\left(\frac{1}{2}{\dot{r}_i}^2\right)$$

$${Q_k}^d = \frac{\partial}{\partial \dot{q}_k}\sum_i\left(-\frac{1}{2}\lambda_i {\dot{r}_i}^2\right)$$

$${Q_k}^d = -\frac{\partial R}{\partial \dot{q}_k} \qquad \text{-------- (5)} \quad [From\ equation\ (2)]$$

Therefore, the Lagrange's equation for any system containing dissipative force is given by

$$\frac{d}{dt}\frac{\partial L}{\partial \dot{q}_k} - \frac{\partial L}{\partial q_k} + \frac{\partial R}{\partial \dot{q}_k} = 0 \qquad \text{------------- (6)}$$

Hence the term $\frac{\partial R}{\partial \dot{q}_k}$ takes into account the dissipative forces.

3.8. A GENERAL EXPRESSION FOR KINETIC ENERGY

Consider a system of N – particles of masses $m_1, m_2, m_3, --- m_N$. If $r_1, r_2, r_3 --- r_i --- r_N$ are the position vectors of particles 1, 2,……, I, ……N respectively, then kinetic energy in terms of Cartesian components of velocity is given by

$$T = \sum_i \frac{1}{2} m_i V_i^2$$

$$T = \sum_i \frac{1}{2} m_i \dot{r}_i^2 \qquad \text{---------------- (1)}$$

We want to express this kinetic energy in terms of generalized co – ordinates $q_1, q_2, q_3 \dots\dots..q_f$ for using in Lagrangian equation
Transformation equation is

$$r_i = r_i(q_1, q_2, q_3 - - - q_k - - - q_f, t) \qquad \text{----------- (2)}$$

Differentiating with respect to time t, we get

$$\dot{r}_i = \sum_k \frac{\partial r_i}{\partial q_k} \dot{q}_k + \frac{\partial r_i}{\partial t} \qquad \text{---------------- (3)}$$

For getting square of such terms we simply multiply the two equivalent terms using different suffixes for summation i.e.,

$$\dot{r}_i^2 = \left(\sum_k \frac{\partial \vec{r}_i}{\partial q_k} \dot{q}_k + \frac{\partial \vec{r}_i}{\partial t}\right) \sum_l \left(\frac{\partial \vec{r}_i}{\partial q_l} \dot{q}_l + \frac{\partial r_i}{\partial t}\right)$$

$$\dot{r}_i^2 = \sum_k \sum_l \frac{\partial \vec{r}_i}{\partial q_k} \frac{\partial \vec{r}_i}{\partial q_l} \dot{q}_k \dot{q}_l + 2 \sum_k \frac{\partial r_i}{\partial q_k} \frac{\partial r_i}{\partial q_t} \dot{q}_k + \left(\frac{\partial r_i}{\partial t}\right)^2 \qquad \text{--------- (4)}$$

Substituting this value in equation (1), we get

$$T = \sum_i \frac{1}{2} m_i \left\{\sum_k \sum_l \frac{\partial \vec{r}_i}{\partial q_k} \frac{\partial \vec{r}_i}{\partial q_l} \dot{q}_k \dot{q}_l + 2 \sum_k \frac{\partial r_i}{\partial q_k} \frac{\partial r_i}{\partial q_t} \dot{q}_k + \left(\frac{\partial r_i}{\partial t}\right)^2 \right\} \qquad \text{--------- (5)}$$

$$T = \sum_k \sum_l a_{kl} \dot{q}_k \dot{q}_l + \sum_k b_k \dot{q}_k + C$$

Where

$$\left.\begin{aligned} a_{kl} &= \sum_i \frac{1}{2} m_i \frac{\partial \vec{r}_i}{\partial q_k} \frac{\partial \vec{r}_i}{\partial q_l} \\ b_k &= \sum_i m_i \frac{\partial r_i}{\partial q_k} \frac{\partial r_i}{\partial q_t} \\ C &= \sum_i \frac{1}{2} m_i \left(\frac{\partial r_i}{\partial t}\right)^2 \end{aligned}\right\} \qquad \text{-------------- (6)}$$

Thus the kinetic energy of a system of particles may be expressed as a sum of three terms
(i) Containing quadratic in generalized velocity
(ii) Containing linear in generalized velocity and
(iii) Independent of generalized velocity.

In case the transformation equation (2) does not contain time, the expression for kinetic energy is of particular interest. The constraints are then scleronomic and the partial derivatives with respect to time disappear. Here $b_k = 0$ and $C = 0$. In this case the kinetic energy term is a homogeneous quadratic function of generalized velocities. Thus

$$T = \sum_k \sum_l a_{kl} \dot{q}_k \dot{q}_l \qquad \text{---------------} (7)$$

3.9. LAGRANGIAN OF A SIMPLE PENDULUM

A simple pendulum consists of a point mass called bob, suspended at the lower end of a massless & inextensible string of constant length (l) fixed at its upper end to a fixed rigid support.

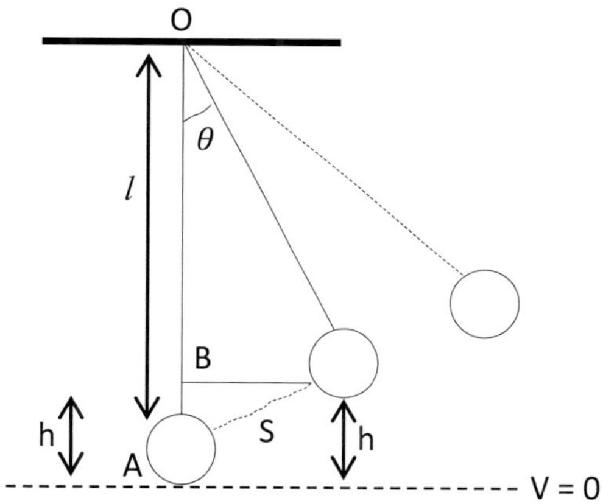

***Fig. (3.10)**: Motion of Simple Pendulum*

Let θ be the angle made by bob at any instant time 't' with its mean position OA

Kinetic Energy is given by

$$T = \frac{1}{2} mV^2 \qquad \text{-----------------} (1)$$

$$\text{Angle} = \frac{Arc}{Radius}$$

$$\theta = \frac{S}{l}$$

$$S = l\theta \qquad \text{-----------------} (2)$$

Differentiating on both sides with respect to time 't', we get

$$\frac{dS}{dt} = l\,\frac{d\theta}{dt}$$

$$V = l\,\dot{\theta} \qquad \text{------------------} (3)$$

Put the value of equation (3) in equation (1), we get

$$T = \frac{1}{2}m\left(l\,\dot{\theta}\right)^2$$

$$T = \frac{1}{2}ml^2\dot{\theta}^2 \quad \text{---------------- (4)}$$

Potential Energy is given by

$$V = mgh$$

$$V = mg(AB)$$

$$V = mg(OA - OB)$$

$$V = mg(l - lcos\theta) \qquad \begin{bmatrix} From\ \Delta OBC \\ Cos\theta = \frac{OB}{OC} \\ Cos\theta = \frac{OB}{l} \\ OB = l\,cos\theta \end{bmatrix}$$

$$V = mgl(1 - cos\theta) \quad \text{---------------- (5)}$$

Lagrangian (L) of the system is given by

$$\text{L = T – V} \quad \text{--------------- (6)}$$

Put the values of equation (4) & equation (5) in equation (6), we get

$$L = \frac{1}{2}ml^2\dot{\theta}^2 - mgl(1 - cos\theta) \quad \text{-------------- (7)}$$

Now, Lagrangian equation of motion for simple pendulum is given by

$$\frac{d}{dt}\left(\frac{\partial L}{\partial \dot{\theta}}\right) - \frac{\partial L}{\partial \theta} = 0 \quad \text{----------------- (8)}$$

Partially differentiating equation (7) with respect to $\dot{\theta}$

$$\frac{\partial L}{\partial \dot{\theta}} = \frac{\partial}{\partial \dot{\theta}}\left(\frac{1}{2}ml^2\dot{\theta}^2 - mgl(1 - cos\theta)\right)$$

$$\frac{\partial L}{\partial \dot{\theta}} = \frac{1}{2}\,m\,l^2\,2\,\dot{\theta} - 0$$

$$\frac{\partial L}{\partial \dot{\theta}} = m\,l^2\,\dot{\theta} \quad \text{---------------- (9)}$$

Partially differentiating equation (7) with respect to θ

$$\frac{\partial L}{\partial \theta} = 0 - mgl\,[0 - (-Sin\theta)]$$

$$\frac{\partial L}{\partial \theta} = -mgl\,Sin\theta \qquad \text{---------------- (10)}$$

Put the values of equation (9) & (10) in equation (8), we get

$$\frac{d}{dt}\left(m\,l^2\,\dot{\theta}\right) - (-mgl\,Sin\theta) = 0$$

$$ml^2\ddot{\theta} + mgl\,Sin\theta = 0$$

Dividing both sides by $m\,l^2$, we get

$$\frac{ml^2\ddot{\theta}}{ml^2} + \frac{mgl\,Sin\theta}{ml^2} = \frac{0}{ml^2}$$

$$\ddot{\theta} + \frac{g}{l}\,Sin\theta = 0$$

For small θ, $\quad Sin\theta \approx \theta$

$$\boldsymbol{\ddot{\theta} + \frac{g}{l}\theta = 0} \qquad \text{---------------- (a)}$$

This is the required Lagrangian equation of motion for a Simple Pendulum

The general equation of Simple Harmonic motion is given by

$$\ddot{\theta} + \omega^2\theta = 0 \qquad \text{---------------- (b)}$$

Comparing equation (a) & (b), we get

$$\omega^2 = \frac{g}{l}$$

$$\omega = \sqrt{\frac{g}{l}}$$

$$2\pi\vartheta = \sqrt{\frac{g}{l}}$$

$Fewquency \quad \vartheta = \frac{1}{2\pi}\sqrt{\frac{g}{l}}$

As $\quad T = \frac{1}{\vartheta}$

$$T = \frac{1}{\frac{1}{2\pi}\sqrt{\frac{g}{l}}}$$

$$T = 2\pi \sqrt{\frac{l}{g}}$$

3.10. LAGRANGIAN OF A SPHERICAL PENDULUM

The spherical pendulum is similar to the simple pendulum but moves in 3 – Dimensional space. In a spherical pendulum, a particle of mass 'm' moves on the surface of sphere of radius 'r' in a gravitational field.

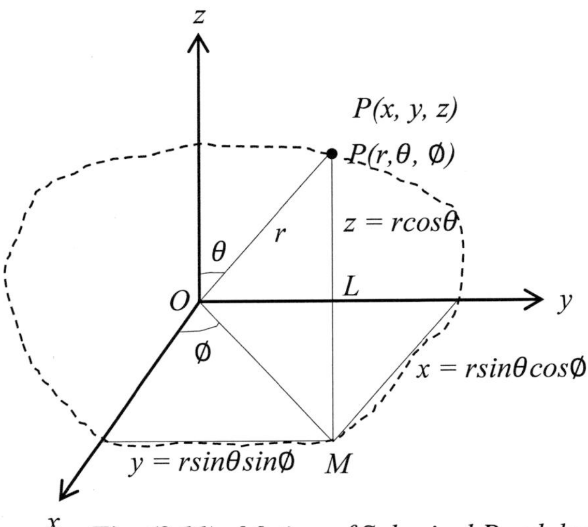

***Fig. (3.11)**: Motion of Spherical Pendulum*

Let $P(x, y, z)$ be the position co – ordinates of the particle moving on the surface of sphere of radius 'r'. Let $(r, \theta, \emptyset)$ are its spherical co – ordinates, then

$$x = r\sin\theta\cos\emptyset \quad \text{--------------} \quad (1)$$

$$y = r\sin\theta\sin\emptyset \quad \text{--------------} \quad (2)$$

$$z = r\cos\theta \quad \text{--------------} \quad (3)$$

Here 'r' is constant and $(\theta, \emptyset)$ are generalized co – ordinates.

The Kinetic Energy (K.E) of the particle is given by

$$T = \frac{1}{2}mV^2$$

$$T = \frac{1}{2}m\left(\frac{dr}{dt}\right)^2$$

$$T = \frac{1}{2}m\dot{r}^2$$

$$T = \frac{1}{2} m \left(\dot{x}^2 + \dot{y}^2 + \dot{z}^2\right) \qquad \text{------------- (4)}$$

From equation (1), we have

$$x = r sin\theta cos\emptyset \text{------------- (1)}$$

Differentiating on both sides, we get

$$\dot{x} = r \left[Sin\theta \left(-Sin\emptyset\right). \dot{\emptyset} + Cos\emptyset \; Cos\theta . \dot{\theta}\right]$$

$$\dot{x} = r \left[-Sin\theta \; Sin\emptyset . \dot{\emptyset} + Cos\emptyset \; Cos\theta . \dot{\theta}\right]$$

$$\dot{x} = r \left[Cos\emptyset \; Cos\theta . \dot{\theta} \; - Sin\theta \; Sin\emptyset . \dot{\emptyset}\right]$$

$$\dot{x} = r \left[\dot{\theta} Cos\emptyset \; Cos\theta \; - \dot{\emptyset} \; Sin\theta \; Sin\emptyset\right]$$

Squaring on both sides, we get

$$\dot{x}^2 = r^2 \left[\dot{\theta} Cos\emptyset \; Cos\theta \; - \dot{\emptyset} \; Sin\theta \; Sin\emptyset\right]^2$$

$$\dot{x}^2 = r^2 \left[\dot{\theta}^2 cos^2\theta cos^2\emptyset + \dot{\emptyset}^2 sin^2\theta sin^2\emptyset - 2\dot{\theta} \; cos\theta cos\emptyset . \dot{\emptyset} \; sin\theta sin\emptyset\right] \qquad \text{------------------ (5)}$$

From equation (2), we have

$$y = r sin\theta sin\emptyset \qquad \text{-------------- (2)}$$

Differentiating on both sides, we get

$$\dot{y} = r \left[sin\theta \; cos\emptyset \; . \dot{\emptyset} + cos\theta \; . \dot{\theta} \; . sin\emptyset\right]$$

Squaring on both sides, we get

$$\dot{y}^2 = r^2 \left[sin\theta \; cos\emptyset \; . \dot{\emptyset} + cos\theta \; . \dot{\theta} \; . sin\emptyset\right]^2$$

$$\dot{y}^2 = r^2 \left[sin^2\theta cos^2\emptyset\left(\dot{\emptyset}^2\right) + cos^2\theta\left(\dot{\theta}^2\right) sin^2\emptyset + 2 sin\theta \; cos\emptyset\left(\dot{\emptyset}\right) cos\theta\left(\dot{\theta}\right) sin\emptyset\right] \qquad \text{------------------ (6)}$$

From equation (3), we have

$$z = r cos\theta \qquad \text{-------------- (3)}$$

$$\dot{z} = -r sin\theta . \dot{\theta}$$

$$\dot{z}^2 = r^2 \; sin^2\theta\left(\dot{\theta}^2\right) \qquad \text{--------------- (7)}$$

Now adding equations (5), (6) & (7) side wise, we get

$$\dot{x}^2 + \dot{y}^2 + \dot{z}^2 = r^2 \left[\dot{\theta}^2 cos^2\theta cos^2\emptyset + \dot{\emptyset}^2 sin^2\theta sin^2\emptyset - 2\dot{\theta}\, cos\theta cos\emptyset . \dot{\emptyset}\, sin\theta sin\emptyset\right] + r^2 \left[sin^2\theta cos^2\emptyset(\dot{\emptyset}^2) + cos^2\theta(\dot{\theta}^2)sin^2\emptyset + 2sin\theta\, cos\emptyset(\dot{\emptyset})\, cos\theta(\dot{\theta})\, sin\emptyset\right] + r^2\, sin^2\theta(\dot{\theta}^2)$$

$$\dot{x}^2 + \dot{y}^2 + \dot{z}^2 = r^2 \left[\dot{\theta}^2 cos^2\theta cos^2\emptyset + \dot{\emptyset}^2 sin^2\theta sin^2\emptyset - 2\dot{\theta}\, cos\theta cos\emptyset . \dot{\emptyset}\, sin\theta sin\emptyset + sin^2\theta cos^2\emptyset(\dot{\emptyset}^2) + cos^2\theta(\dot{\theta}^2)sin^2\emptyset + 2sin\theta\, cos\emptyset(\dot{\emptyset})\, cos\theta(\dot{\theta})\, sin\emptyset + sin^2\theta(\dot{\theta}^2)\right]$$

$$\dot{x}^2 + \dot{y}^2 + \dot{z}^2 = r^2 \left[\dot{\theta}^2 cos^2\theta cos^2\emptyset + \dot{\emptyset}^2 sin^2\theta sin^2\emptyset + sin^2\theta cos^2\emptyset(\dot{\emptyset}^2) + cos^2\theta(\dot{\theta}^2)sin^2\emptyset + sin^2\theta(\dot{\theta}^2)\right]$$

$$\dot{x}^2 + \dot{y}^2 + \dot{z}^2 = r^2 \left[\dot{\theta}^2 cos^2\theta cos^2\emptyset + cos^2\theta(\dot{\theta}^2)sin^2\emptyset + \dot{\emptyset}^2 sin^2\theta sin^2\emptyset + sin^2\theta cos^2\emptyset(\dot{\emptyset}^2) + sin^2\theta(\dot{\theta}^2)\right]$$

$$\dot{x}^2 + \dot{y}^2 + \dot{z}^2 = r^2 \left[\dot{\theta}^2 cos^2\theta\, (cos^2\emptyset + sin^2\emptyset) + \dot{\emptyset}^2 sin^2\theta\, (sin^2\emptyset + cos^2\emptyset) + \dot{\theta}^2 sin^2\theta\right]$$

$$\dot{x}^2 + \dot{y}^2 + \dot{z}^2 = r^2 \left[\dot{\theta}^2 cos^2\theta\, (1) + \dot{\emptyset}^2 sin^2\theta\, (1) + \dot{\theta}^2 sin^2\theta\right]$$

$$\dot{x}^2 + \dot{y}^2 + \dot{z}^2 = r^2 \left[\dot{\theta}^2 cos^2\theta + \dot{\emptyset}^2 sin^2\theta + \dot{\theta}^2 sin^2\theta\right]$$

$$\dot{x}^2 + \dot{y}^2 + \dot{z}^2 = r^2 \left[\dot{\theta}^2\, (cos^2\theta + sin^2\theta) + \dot{\emptyset}^2 sin^2\theta\right]$$

$$\dot{x}^2 + \dot{y}^2 + \dot{z}^2 = r^2 \left[\dot{\theta}^2\, (1) + \dot{\emptyset}^2 sin^2\theta\right]$$

$$\dot{x}^2 + \dot{y}^2 + \dot{z}^2 = r^2 \left[\dot{\theta}^2 + \dot{\emptyset}^2 sin^2\theta\right] \quad \text{----------------- (8)}$$

Now put the value of equation (8) in equation (4), we get

$$T = \frac{1}{2} m\, (\dot{x}^2 + \dot{y}^2 + \dot{z}^2) \quad \text{------------ (4)}$$

$$T = \frac{1}{2} m\, r^2 \left(\dot{\theta}^2 + \dot{\emptyset}^2 sin^2\theta\right) \quad \text{------------- (9)}$$

Potential Energy (V) of the particle is given by

$$V = mgh$$

$$V = mgz$$

$$V = mgrcos\theta \quad \text{--------------- (10)}$$

Lagrangian (L) of the system is given by

$$L = T - V \qquad \text{-----------------} (11)$$

Put the values of equations (9) & (10) in equation (11), we get

$$L = \frac{1}{2} m\, r^2 \left(\dot{\theta}^2 + \dot{\emptyset}^2 sin^2\theta\right) - mgrcos\theta \quad \text{------------} (12)$$

Now partially differentiating equation (12) with respect to $\dot{\theta}$, we get

$$\frac{\partial L}{\partial \dot{\theta}} = \frac{\partial}{\partial \dot{\theta}} \left\{ \frac{1}{2} m\, r^2 \left(\dot{\theta}^2 + \dot{\emptyset}^2 sin^2\theta\right) - mgrcos\theta \right\}$$

$$\frac{\partial L}{\partial \dot{\theta}} = \frac{1}{2} mr^2 2\dot{\theta}$$

$$\frac{\partial L}{\partial \dot{\theta}} = mr^2\dot{\theta} \qquad \text{-----------------} (13)$$

Now again partially differentiating equation (12) w.r.to , we get

$$\frac{\partial L}{\partial \theta} = \frac{\partial}{\partial \theta} \left\{ \frac{1}{2} m\, r^2 \left(\dot{\theta}^2 + \dot{\emptyset}^2 sin^2\theta\right) - mgrcos\theta \right\}$$

$$\frac{\partial L}{\partial \theta} = \frac{1}{2} mr^2 \left[0 + \dot{\emptyset}^2 2sin\theta(-cos\theta)\right] + mgrsin\theta$$

$$\frac{\partial L}{\partial \theta} = mr^2\dot{\emptyset}^2 sin\theta cos\theta + mgrsin\theta \quad \text{--------------} (14)$$

Again Partially differentiating equation (12) with respect to $\dot{\emptyset}$, we get

$$\frac{\partial L}{\partial \dot{\emptyset}} = \frac{\partial}{\partial \dot{\emptyset}} \left\{ \frac{1}{2} m\, r^2 \left(\dot{\theta}^2 + \dot{\emptyset}^2 sin^2\theta\right) - mgrcos\theta \right\}$$

$$\frac{\partial L}{\partial \dot{\emptyset}} = \frac{1}{2} mr^2 \left[0 + 2\dot{\emptyset} sin^2\theta \,\right] - 0$$

$$\frac{\partial L}{\partial \dot{\emptyset}} = mr^2\, \dot{\emptyset} sin^2\theta \qquad \text{-----------------} (15)$$

Again Partially differentiating equation (12) with respect to $\emptyset$, we get

$$\frac{\partial L}{\partial \emptyset} = \frac{\partial}{\partial \emptyset} \left\{ \frac{1}{2} m\, r^2 \left(\dot{\theta}^2 + \dot{\emptyset}^2 sin^2\theta\right) - mgrcos\theta \right\}$$

$$\frac{\partial L}{\partial \emptyset} = 0 \qquad \text{-----------------} (16)$$

Lagrangian equation of motion for a spherical pendulum s given by

In terms of θ (Here θ is generalized Co – ordinate)

$$\frac{d}{dt}\left(\frac{\partial L}{\partial \dot{\theta}}\right) - \frac{\partial L}{\partial \theta} = 0 \quad \text{------------------- (17)}$$

In terms of ∅ (Here ∅ is generalized Co – ordinate)

$$\frac{d}{dt}\left(\frac{\partial L}{\partial \dot{\emptyset}}\right) - \frac{\partial L}{\partial \emptyset} = 0 \quad \text{------------------- (18)}$$

Put the values of equation (13) & (14) in equation (17), we get

$$\frac{d}{dt}\left(mr^2\dot{\theta}\right) - \left(mr^2\dot{\emptyset}^2 sin\theta cos\theta + mgrsin\theta\right) = 0$$

$$mr^2\ddot{\theta} - mr^2\dot{\emptyset}^2 sin\theta cos\theta - mgrsin\theta = 0$$

Dividing on both sides with mr^2, we get

$$\frac{mr^2\ddot{\theta}}{mr^2} - \frac{mr^2\dot{\emptyset}^2 sin\theta cos\theta}{mr^2} - \frac{mgrsin\theta}{mr^2} = \frac{0}{mr^2}$$

$$\ddot{\theta} - \dot{\emptyset}^2 sin\theta cos\theta - \frac{g}{r}sin\theta = 0 \quad \text{----------------- (19)}$$

Put the values of equation (15) & (16) in equation (18), we get

$$\frac{d}{dt}\left(mr^2\,\dot{\emptyset}sin^2\theta\right) - 0 = 0$$

Integrating on both sides, we get

$$mr^2\,\dot{\emptyset}sin^2\theta = P_{\emptyset}$$

$$\dot{\emptyset} = \frac{P_{\emptyset}}{mr^2 sin^2\theta} \quad \text{----------------- (20)}$$

Where, $P_{\emptyset}$ - is constant of angular momentum

Put the values of equation (20) in equation (19), we get

$$\ddot{\theta} - \left(\frac{P_{\emptyset}}{mr^2 sin^2\theta}\right)^2 sin\theta cos\theta - \frac{g}{r}sin\theta = 0$$

$$\ddot{\theta} - \frac{{P_{\emptyset}}^2}{m^2r^4 sin^4\theta} sin\theta cos\theta - \frac{g}{r}sin\theta = 0$$

$$\ddot{\theta} - \frac{P_\emptyset{}^2}{m^2r^4sin^3\theta}cos\theta - \frac{g}{r}sin\theta = 0$$

$$\boldsymbol{\ddot{\theta} - \frac{P_\emptyset{}^2 cos\theta}{m^2r^4sin^3\theta} - \frac{g}{r}sin\theta = 0}$$

This is the required equation of motion for spherical pendulum.

3.11. LAGRANGIAN OF ATWOOD'S MACHINE

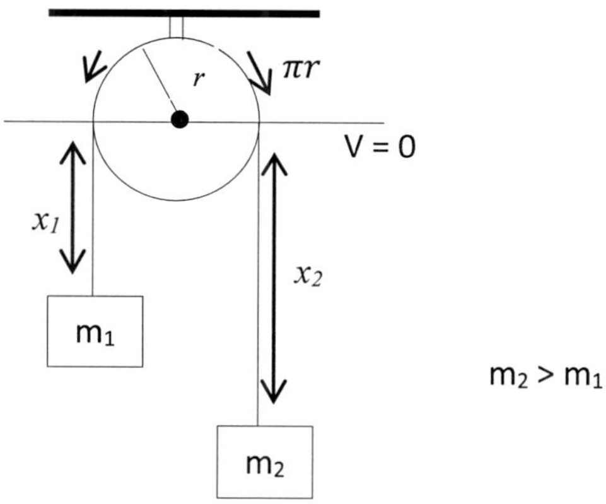

Fig. (3.12): *Atwood's Machine*

Atwood's machine consists of 2 masses m_1 & m_2 suspended over a frictionless pulley of radius *'r'* and connected by an inextensible string of constant length *'l'* as shown in Fig (1.8). Since the masses can move only in vertical direction. This is one – dimensional problem.

As $m_2 > m_1$, velocity of $\frac{dx}{dt}$ of m_2 will be downwards, whereas velocity of m_1 will be upwards we can neglect the friction between the string and pulley. The acting on m_1 moving upwards.

From figure (1.8) it is clear that

$$l = x_1 + \pi r + x_2$$

$$x_2 = l - x_1 - \pi r \qquad \text{-------------- (1)}$$

Differentiate equation (1) w.r.to. t

$$\dot{x}_2 = 0 - \dot{x}_1 - 0$$

$$\dot{x}_2 = - \dot{x}_1 \qquad \text{-------------- (2)}$$

Kinetic Energy (K.E) of mass m_1 is given by

$$T_1 = \frac{1}{2} m_1 V_1^2$$

$$T_1 = \frac{1}{2} m_1 \dot{x}_1^2 \quad \text{----------} (3) \left[V_1 = \frac{dx_1}{dt} = \dot{x}_1\right]$$

Kinetic Energy (K.E) of mass m_2 is given by

$$T_2 = \frac{1}{2} m_2 V_2^2$$

$$T_2 = \frac{1}{2} m_2 \dot{x}_2^2 \quad \text{----------} (4) \left[V_2 = \frac{dx_2}{dt} = \dot{x}_2\right]$$

Using Equation (2) in Equation (1), we get

$$T_2 = \frac{1}{2} m_2(-\dot{x}_1)^2$$

$$T_2 = \frac{1}{2} m_2 \dot{x}_1^2 \quad \text{-----------------} (5)$$

Total kinetic Energy (K.E) of the system is given by

$$T = T_1 + T_2$$

$$T = \frac{1}{2} m_1 \dot{x}_1^2 + \frac{1}{2} m_2 \dot{x}_1^2$$

$$T = \frac{1}{2} (m_1 + m_2) \dot{x}_1^2 \quad \text{---------------} (6)$$

Potential Energy (P.E) of mass m_1 is given by

$$V_1 = -m_1 g x_1 \quad \text{---------------} (7) \quad [P.E = -mgh]$$

Potential Energy (P.E) of mass m_1 is given by

$$V_2 = -m_2 g x_2 \quad \text{---------------} (8) \quad [P.E = -mgh]$$

Using Equation (1) in Equation (8), we get

$$V_2 = -m_2 g\,(l - x_1 - \pi r) \quad \text{--------------} (9)$$

∴ Total P.E of the system is given by

$$V = V_1 + V_2$$

$$V = -m_1 g x_1 - m_2 g\,(l - x_1 - \pi r)$$

$$V = -m_1 g x_1 - m_2 g l + m_2 g\, x_1 + m_2 g \pi r$$

$$V = -m_1 g x_1 + m_2 g\, x_1 - m_2 g l + m_2 g \pi r$$

$$V = -g\, x_1 (m_1 - m_2) - m_2 g\,(l - \pi r)$$

$$V = -g\,x_1\,(m_1 - m_2) + V_0 \qquad \text{--------------- (10)}$$

Where $V_0 = -\,m_2 g\,(l - \pi r)$ = Constant

Lagrangian L of the system is given by

$$L = T - V \qquad \text{--------------- (11)}$$

Using Equation (6) & (10) in Equation (11), we get

$$L = \frac{1}{2}\,(m_1 + m_2)\,\dot{x}_1^{\,2} - [-\,g\,x_1\,(m_1 - m_2) + V_0]$$

$$L = \frac{1}{2}\,(m_1 + m_2)\,\dot{x}_1^{\,2} + g\,x_1\,(m_1 - m_2) + V_0 \qquad \text{------- (12)}$$

Now Lagrange's Equation of motion is given by

$$\frac{d}{dt}\left(\frac{\partial L}{\partial \dot{x}_1}\right) - \frac{\partial L}{\partial x_1} = 0 \qquad \text{--------------------- (13)}$$

Partially differentiating Equation (12) w.r.to. $\dot{x}_1$ we get

$$\frac{\partial L}{\partial \dot{x}_1} = \frac{1}{2}\,(m_1 + m_2)\,2\,\dot{x}_1 + 0$$

$$\frac{\partial L}{\partial \dot{x}_1} = (m_1 + m_2)\,\dot{x}_1 \qquad \text{----------------- (14)}$$

Again, partially differentiating Equation (12) w.r.to. x_1 we get

$$\frac{\partial L}{\partial x_1} = 0 + g\,(m_1 - m_2)$$

$$\frac{\partial L}{\partial x_1} = g\,(m_1 - m_2) \qquad \text{----------------- (15)}$$

Using Equation (14) & (15) in Equation (13), we get

$$\frac{d}{dt}\,(m_1 + m_2)\,\dot{x}_1 - g\,(m_1 - m_2) = 0$$

$$(m_1 + m_2)\,\ddot{x}_1 - g\,(m_1 - m_2) = 0$$

$$(m_1 + m_2)\,\ddot{x}_1 = g\,(m_1 - m_2)$$

$$\ddot{x}_1 = \frac{g\,(m_1 - m_2)}{m_1 + m_2} \qquad \text{-------------------- (16)}$$

3.12. LAGRANGIAN OF A DOUBLE PENDULUM

A double pendulum consists of 2 masses m_1 & m_2 oscillating in the same plane. First bob of mass 'm_1' is suspended by an inextensible rod of length 'l_1' at point 'O' and the other bob of mass 'm_2'

suspended from another light rod of length 'l_2' which is connected with first bob 'm_1'. Let θ_1 & θ_2 be the deflections of the pendulum from the vertical. Let (x_1, y_1) & (x_2, y_2) be the rectangular co – ordinates of masses m_1 & m_2 respectively at time 't'.

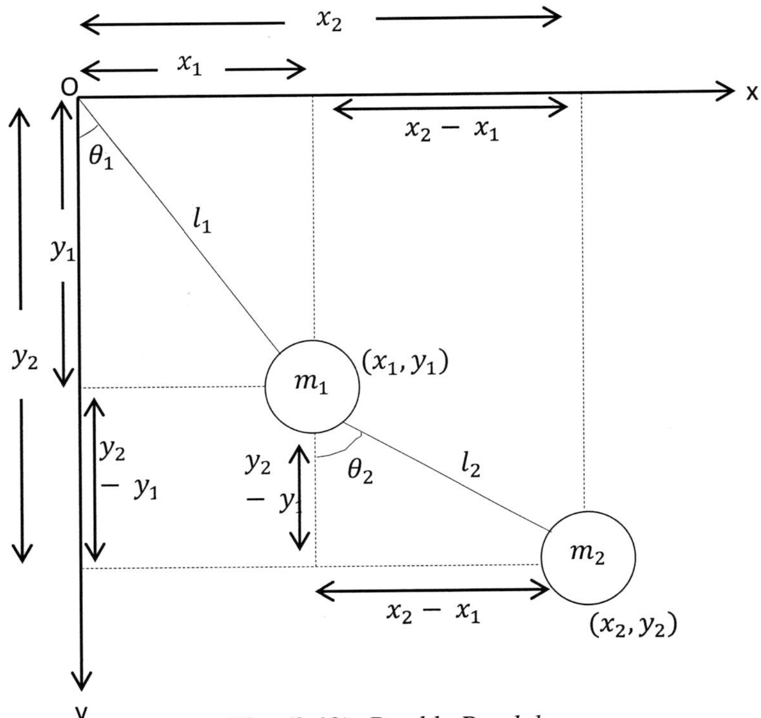

Fig. (3.13): Double Pendulum

From Figure, we have $Sin\theta_1 = \frac{x_1}{l_1}$

$$x_1 = l_1 \, Sin\theta_1 \quad \text{---------------- (1)}$$

$$\dot{x_1} = l_1 Cos\theta_1 . \dot{\theta_1} \quad \text{---------------- (2)}$$

Also, $Cos\theta_1 = \frac{y_1}{l_1}$

$$y_1 = l_1 Cos\theta_1 \quad \text{---------------- (3)}$$

$$\dot{y_1} = -l_1 Sin\theta_1 . \dot{\theta_1} \text{---------------- (4)}$$

Also, $Sin\theta_2 = \frac{x_2 - x_1}{l_2}$

$$x_2 - x_1 = l_2 Sin\theta_2$$

$$x_2 = x_1 + l_2 Sin\theta_2$$

$$x_2 = l_1\, Sin\theta_1 + l_2 Sin\theta_2 \quad \text{---------} \ (5) \begin{bmatrix} We\ know\ that \\ x_1 = l_1\, Sin\theta_1 \end{bmatrix}$$

$$\dot{x}_2 = l_1 Cos\theta_1.\dot{\theta}_1 + l_2 Cos\theta_2.\dot{\theta}_2 \quad \text{-----------} \ (6)$$

Also, $$Cos\theta_2 = \frac{y_2 - y_1}{l_2}$$

$$y_2 - y_1 = l_2 Cos\theta_2$$

$$y_2 = y_1 + l_2 Cos\theta_2$$

$$y_2 = l_1\, Cos\theta_1 + l_2 Cos\theta_2 \quad \text{---------} \ (7) \begin{bmatrix} We\ know\ that \\ y_1 = l_1 Cos\theta_1 \end{bmatrix}$$

$$\dot{y}_2 = -l_1 Sin\theta_1.\dot{\theta}_1 - l_2 Sin\theta_2.\dot{\theta}_2$$

$$\dot{y}_2 = -(l_1 Sin\theta_1.\dot{\theta}_1 + l_2 Sin\theta_2.\dot{\theta}_2) \quad \text{-------------} \ (8)$$

Kinetic Energy T = K.E of mass m_1 + K.E of mass m_2

$$T = \frac{1}{2} m_1 (\dot{x}_1^{\,2} + \dot{y}_1^{\,2}) + \frac{1}{2} m_2 (\dot{x}_2^{\,2} + \dot{y}_2^{\,2}) \quad \text{--------------} \ (9)$$

Now put the values of equations (2), (4), (6) & (8) in equation (9), we get

$$T = \frac{1}{2} m_1 \left[(l_1 Cos\theta_1.\dot{\theta}_1)^2 + (-l_1 Sin\theta_1.\dot{\theta}_1)^2 \right] + \frac{1}{2} m_2 \left[(l_1 Cos\theta_1.\dot{\theta}_1 + l_2 Cos\theta_2.\dot{\theta}_2)^2 + (-(l_1 Sin\theta_1.\dot{\theta}_1 + l_2 Sin\theta_2.\dot{\theta}_2))^2 \right]$$

$$T = \frac{1}{2} m_1 \left[l_1^{\,2} Cos^2\theta_1 \dot{\theta}_1^{\,2} + l_1^{\,2} Sin^2\theta_1 \dot{\theta}_1^{\,2} \right] + \frac{1}{2} m_2 \left[(l_1 Cos\theta_1.\dot{\theta}_1 + l_2 Cos\theta_2.\dot{\theta}_2)^2 + (l_1 Sin\theta_1.\dot{\theta}_1 + l_2 Sin\theta_2.\dot{\theta}_2)^2 \right]$$

$$T = \frac{1}{2} m_1 \left[l_1^{\,2} \dot{\theta}_1^{\,2} (Sin^2\theta_1 + Cos^2\theta_1) \right] + \frac{1}{2} m_2 \Big[\big(l_1^{\,2} Cos^2\theta_1 \dot{\theta}_1^{\,2} + l_2^{\,2} Cos^2\theta_2 \dot{\theta}_2^{\,2} + 2 l_1 Cos\theta_1.\dot{\theta}_1 l_2 Cos\theta_2.\dot{\theta}_2 \big) + \big(l_1^{\,2} Sin^2\theta_1 \dot{\theta}_1^{\,2} + l_2^{\,2} Sin^2\theta_2 \dot{\theta}_2^{\,2} + 2 l_1 Sin\theta_1.\dot{\theta}_1 l_2 Sin\theta_2.\dot{\theta}_2 \big) \Big]$$

$$T = \frac{1}{2}m_1\left[l_1{}^2\dot{\theta}_1{}^2\,(1)\right] + \frac{1}{2}m_2\left[l_1{}^2\dot{\theta}_1{}^2(Cos^2\theta_1 + Sin^2\theta_1\,) + \,l_2{}^2\dot{\theta}_2{}^2(Cos^2\theta_2 + Sin^2\theta_2) + 2l_1l_2\dot{\theta}_1\dot{\theta}_2\,(Cos\theta_1Cos\theta_2 + Sin\theta_1Sin\theta_2)\right]$$

$$T = \frac{1}{2}m_1\left[l_1{}^2\dot{\theta}_1{}^2\,(1)\right] + \frac{1}{2}m_2\left[l_1{}^2\dot{\theta}_1{}^2(1\,) + \,l_2{}^2\dot{\theta}_2{}^2(1) + 2l_1l_2\dot{\theta}_1\dot{\theta}_2\,(Cos\theta_1Cos\theta_2 + Sin\theta_1Sin\theta_2)\right]$$

$$T = \frac{1}{2}m_1\left[l_1{}^2\dot{\theta}_1{}^2\right] + \frac{1}{2}m_2\left[l_1{}^2\dot{\theta}_1{}^2 + \,l_2{}^2\dot{\theta}_2{}^2 + 2l_1l_2\dot{\theta}_1\dot{\theta}_2\,Cos(\theta_1 - \theta_2)\right] \quad \text{---------} \; (10)$$

Taking reference level as the horizontal plane passing through the point of suspension 'O'

Potential Energy (V) = P.E of mass m_1 + P.E of mass m_2

$$V = -m_1gy_1 - \,m_2gy_2 \qquad \text{---------------} \; (11)$$

Put the values of equations (3) & (7) in equation (11), we get

$$V = -m_1g(l_1Cos\theta_1) - \,m_2g(l_1\,Cos\theta_1 + \,l_2Cos\theta_2)$$

$$V = -m_1gl_1Cos\theta_1 - \,m_2gl_1\,Cos\theta_1 - \,m_2gl_2Cos\theta_2$$

$$V = -\,(m_1 + m_2)gl_1Cos\theta_1 - \,m_2gl_2Cos\theta_2 \quad \text{-------------} \; (12)$$

Lagrangian 'L' of the system is given by

$$L = T - V \qquad \text{----------------} \; (13)$$

Now put the values of equations (10) & (12) in equation (13), we get

$$L = \frac{1}{2}m_1\left[l_1{}^2\dot{\theta}_1{}^2\right] + \frac{1}{2}m_2\left[l_1{}^2\dot{\theta}_1{}^2 + \,l_2{}^2\dot{\theta}_2{}^2 + 2l_1l_2\dot{\theta}_1\dot{\theta}_2\,Cos(\theta_1 - \theta_2)\right] - [-\,(m_1 + m_2)gl_1Cos\theta_1 - \,m_2gl_2Cos\theta_2]$$

$$L = \frac{1}{2}m_1\left[l_1{}^2\dot{\theta}_1{}^2\right] + \frac{1}{2}m_2\left[l_1{}^2\dot{\theta}_1{}^2 + \,l_2{}^2\dot{\theta}_2{}^2 + 2l_1l_2\dot{\theta}_1\dot{\theta}_2\,Cos(\theta_1 - \theta_2)\right] + (m_1 + m_2)gl_1Cos\theta_1 + m_2gl_2Cos\theta_2 \quad \text{--------------} \; (14)$$

Lagrange's equation of motion for generalized co – ordinates 'θ_1' is given by

$$\frac{d}{dt}\left(\frac{\partial L}{\partial \dot{\theta}_1}\right) - \frac{\partial L}{\partial \theta_1} = 0 \qquad \text{----------------} \; (15)$$

Now partially differentiating equation (14) w. r. to. $\dot{\theta}_1$, we get

$$\frac{\partial L}{\partial \dot{\theta}_1} = \frac{\partial}{\partial \dot{\theta}_1}\left\{\frac{1}{2}m_1\left[{l_1}^2{\dot{\theta}_1}^2\right] + \frac{1}{2}m_2\left[{l_1}^2{\dot{\theta}_1}^2 + {l_2}^2{\dot{\theta}_2}^2 + 2l_1l_2\dot{\theta}_1\dot{\theta}_2\,Cos(\theta_1 - \theta_2)\right] + (m_1+m_2)gl_1Cos\theta_1 + m_2gl_2Cos\theta_2\right\}$$

$$\frac{\partial L}{\partial \dot{\theta}_1} = \frac{1}{2}m_1{l_1}^2 2\dot{\theta}_1 + \frac{1}{2}m_2\left[{l_1}^2 2\dot{\theta}_1 + 0 + 2l_1l_2\dot{\theta}_2\,Cos(\theta_1 - \theta_2) + 0 + 0\right]$$

$$\frac{\partial L}{\partial \dot{\theta}_1} = m_1{l_1}^2\dot{\theta}_1 + \frac{1}{2}m_2\,{l_1}^2 2\dot{\theta}_1 + \frac{1}{2}m_2 2l_1l_2\dot{\theta}_2\,Cos(\theta_1 - \theta_2)$$

$$\frac{\partial L}{\partial \dot{\theta}_1} = m_1{l_1}^2\dot{\theta}_1 + m_2\,{l_1}^2\dot{\theta}_1 + m_2l_1l_2\dot{\theta}_2\,Cos(\theta_1 - \theta_2) \quad \text{------------- (16)}$$

Again partially differentiating equation (14) w. r. to. θ_1 , we get

$$\frac{\partial L}{\partial \theta_1} = \frac{\partial}{\partial \theta_1}\left\{\frac{1}{2}m_1\left[{l_1}^2{\dot{\theta}_1}^2\right] + \frac{1}{2}m_2\left[{l_1}^2{\dot{\theta}_1}^2 + {l_2}^2{\dot{\theta}_2}^2 + 2l_1l_2\dot{\theta}_1\dot{\theta}_2\,Cos(\theta_1 - \theta_2)\right] + (m_1+m_2)gl_1Cos\theta_1 + m_2gl_2Cos\theta_2\right\}$$

$$\frac{\partial L}{\partial \theta_1} = 0 + \frac{1}{2}m_2\left[0 + 0 - 2l_1l_2\dot{\theta}_1\dot{\theta}_2\,Sin(\theta_1 - \theta_2)(1-0)\right] - (m_1+m_2)gl_1Sin\theta_1 + 0$$

$$\frac{\partial L}{\partial \theta_1} = \frac{1}{2}m_2\left[-2l_1l_2\dot{\theta}_1\dot{\theta}_2\,Sin(\theta_1 - \theta_2)(1)\right] - (m_1+m_2)gl_1Sin\theta_1$$

$$\frac{\partial L}{\partial \theta_1} = -m_2l_1l_2\dot{\theta}_1\dot{\theta}_2\,Sin(\theta_1 - \theta_2) - (m_1+m_2)gl_1Sin\theta_1 \quad \text{-------- (17)}$$

Put the values of equation (16) & (17) in equation (15), we get

$$\frac{d}{dt}\left[m_1{l_1}^2\dot{\theta}_1 + m_2\,{l_1}^2\dot{\theta}_1 + m_2l_1l_2\dot{\theta}_2\,Cos(\theta_1 - \theta_2)\right] - \left[-m_2l_1l_2\dot{\theta}_1\dot{\theta}_2\,Sin(\theta_1 - \theta_2) - (m_1+m_2)gl_1Sin\theta_1\right] = 0$$

$$m_1{l_1}^2\ddot{\theta}_1 + m_2\,{l_1}^2\ddot{\theta}_1 + m_2l_1l_2\ddot{\theta}_2\,Cos(\theta_1 - \theta_2) - m_2l_1l_2\dot{\theta}_2\,Sin(\theta_1 - \theta_2)(\dot{\theta}_1 - \dot{\theta}_2) + m_2l_1l_2\dot{\theta}_1\dot{\theta}_2\,Sin(\theta_1 - \theta_2) + (m_1+m_2)gl_1Sin\theta_1 = 0$$

$$(m_1+m_2)\,{l_1}^2\ddot{\theta}_1 + m_2l_1l_2\ddot{\theta}_2\,Cos(\theta_1 - \theta_2) - m_2l_1l_2\dot{\theta}_2\dot{\theta}_1\,Sin(\theta_1 - \theta_2) + m_2l_1l_2{\dot{\theta}_2}^2\,Sin(\theta_1 - \theta_2) + m_2l_1l_2\dot{\theta}_1\dot{\theta}_2\,Sin(\theta_1 - \theta_2) + (m_1+m_2)gl_1Sin\theta_1 = 0$$

$(m_1 + m_2)\, l_1{}^2 \ddot{\theta}_1 + m_2 l_1 l_2 \ddot{\theta}_2\, Cos(\theta_1 - \theta_2) +\ m_2 l_1 l_2 \dot{\theta}_1 \dot{\theta}_2\, Sin(\theta_1 - \theta_2) +$
$(m_1 + m_2) g l_1 Sin\theta_1 = 0$ -------------- (18)

Lagrange's equation of motion for generalized co – ordinates 'θ_2' is given by

$$\frac{d}{dt}\left(\frac{\partial L}{\partial \dot{\theta}_2}\right) - \frac{\partial L}{\partial \theta_2} = 0 \qquad \text{---------------- (19)}$$

Now partially differentiating equation (14) w. r. to. $\dot{\theta}_2$, we get

$$\frac{\partial L}{\partial \dot{\theta}_2} = \frac{\partial}{\partial \dot{\theta}_2}\left\{\frac{1}{2} m_1 \left[l_1{}^2 \dot{\theta}_1{}^2\right] + \frac{1}{2} m_2 \left[l_1{}^2 \dot{\theta}_1{}^2 +\ l_2{}^2 \dot{\theta}_2{}^2 + 2 l_1 l_2 \dot{\theta}_1 \dot{\theta}_2\, Cos(\theta_1 - \theta_2)\right] +\ (m_1 + m_2) g l_1 Cos\theta_1 + m_2 g l_2 Cos\theta_2\right\}$$

$$\frac{\partial L}{\partial \dot{\theta}_2} = 0 + \frac{1}{2} m_2 \left[0 +\ l_2{}^2\, 2\, \dot{\theta}_2 +\ 2 l_1 l_2 \dot{\theta}_1 \dot{\theta}_2\, Cos(\theta_1 - \theta_2)\right] + 0 + 0$$

$$\frac{\partial L}{\partial \dot{\theta}_2} = \frac{1}{2} m_2 \left[l_2{}^2\, 2\, \dot{\theta}_2 +\ 2 l_1 l_2 \dot{\theta}_1 \dot{\theta}_2\, Cos(\theta_1 - \theta_2)\right]$$

$$\frac{\partial L}{\partial \dot{\theta}_2} = \frac{1}{2} m_2 l_2{}^2\, 2\, \dot{\theta}_2 + \frac{1}{2} m_2\, 2 l_1 l_2 \dot{\theta}_1 \dot{\theta}_2\, Cos(\theta_1 - \theta_2)$$

$$\frac{\partial L}{\partial \dot{\theta}_2} = m_2 l_2{}^2\, \dot{\theta}_2 + m_2\, l_1 l_2 \dot{\theta}_1 \dot{\theta}_2\, Cos(\theta_1 - \theta_2) \qquad \text{------------- (20)}$$

Again partially differentiating equation (14) w. r. to. θ_2 , we get

$$\frac{\partial L}{\partial \theta_2} = \frac{\partial}{\partial \theta_2}\left\{\frac{1}{2} m_1 \left[l_1{}^2 \dot{\theta}_1{}^2\right] + \frac{1}{2} m_2 \left[l_1{}^2 \dot{\theta}_1{}^2 +\ l_2{}^2 \dot{\theta}_2{}^2 + 2 l_1 l_2 \dot{\theta}_1 \dot{\theta}_2\, Cos(\theta_1 - \theta_2)\right] + (m_1 + m_2) g l_1 Cos\theta_1 + m_2 g l_2 Cos\theta_2\right\}$$

$$\frac{\partial L}{\partial \theta_2} = 0 + \frac{1}{2} m_2 \left[0 + 0 - 2 l_1 l_2 \dot{\theta}_1 \dot{\theta}_2\, Sin(\theta_1 - \theta_2)(0 - 1)\right] + 0 -\ m_2 g l_2 Sin\theta_2$$

$$\frac{\partial L}{\partial \theta_2} =\ \frac{1}{2} m_2 \left[-2 l_1 l_2 \dot{\theta}_1 \dot{\theta}_2\, Sin(\theta_1 - \theta_2)(-1)\right] -\ m_2 g l_2 Sin\theta_2$$

$$\frac{\partial L}{\partial \theta_2} =\ \frac{1}{2} m_2 2 l_1 l_2 \dot{\theta}_1 \dot{\theta}_2\, Sin(\theta_1 - \theta_2) -\ m_2 g l_2 Sin\theta_2$$

$$\frac{\partial L}{\partial \theta_2} = m_2 l_1 l_2 \dot{\theta}_1 \dot{\theta}_2\, Sin(\theta_1 - \theta_2) -\ m_2 g l_2 Sin\theta_2 \qquad \text{------------- (21)}$$

Now put the values of equations (20) & (21) in equation (19), we get

$$\frac{d}{dt}\left[m_2 l_2^{\,2}\,\dot{\theta}_2 + m_2\, l_1 l_2 \dot{\theta}_1 \dot{\theta}_2\, Cos(\theta_1-\theta_2)\right] - \left[m_2 l_1 l_2 \dot{\theta}_1 \dot{\theta}_2\, Sin(\theta_1-\theta_2) - m_2 g l_2 Sin\theta_2\,\right] = 0$$

$$m_2 l_2^{\,2}\,\ddot{\theta}_2 + m_2\, l_1 l_2 \ddot{\theta}_1 Cos(\theta_1-\theta_2) - m_2\, l_1 l_2 \dot{\theta}_1 Sin(\theta_1-\theta_2)\left(\dot{\theta}_1 - \dot{\theta}_2\right) - m_2 l_1 l_2 \dot{\theta}_1 \dot{\theta}_2\, Sin(\theta_1-\theta_2) + m_2 g l_2 Sin\theta_2 = 0$$

$$m_2 l_2^{\,2}\,\ddot{\theta}_2 + m_2\, l_1 l_2 \ddot{\theta}_1 Cos(\theta_1-\theta_2) - m_2\, l_1 l_2 \dot{\theta}_1^{\,2} Sin(\theta_1-\theta_2) + m_2 l_1 l_2 \dot{\theta}_1 \dot{\theta}_2\, Sin(\theta_1-\theta_2) - m_2 l_1 l_2 \dot{\theta}_1 \dot{\theta}_2\, Sin(\theta_1-\theta_2) + m_2 g l_2 Sin\theta_2 = 0$$

$$m_2 l_2^{\,2}\,\ddot{\theta}_2 + m_2\, l_1 l_2 \ddot{\theta}_1 Cos(\theta_1-\theta_2) - m_2\, l_1 l_2 \dot{\theta}_1^{\,2} Sin(\theta_1-\theta_2) + m_2 g l_2 Sin\theta_2 = 0$$

$$m_2 l_2\left[l_2 \ddot{\theta}_2 + l_1 \ddot{\theta}_1 Cos(\theta_1-\theta_2) - l_1 \dot{\theta}_1^{\,2} Sin(\theta_1-\theta_2) + g Sin\theta_2\right] = 0$$

$$l_2 \ddot{\theta}_2 + l_1 \ddot{\theta}_1 Cos(\theta_1-\theta_2) - l_1 \dot{\theta}_1^{\,2} Sin(\theta_1-\theta_2) + g Sin\theta_2 = 0 \quad \text{-------} \ (22)$$

Equations (18) & (22) are the required Lagrange's equation of motion for Double Pendulum

3.13. L – C CIRCUIT IN LAGRANGIAN FORMULATION

Consider an electrical circuit comprising an inductor of inductance $L^{\prime}$ and Capacitor of capacitance 'C' as shown in figure (3.14).

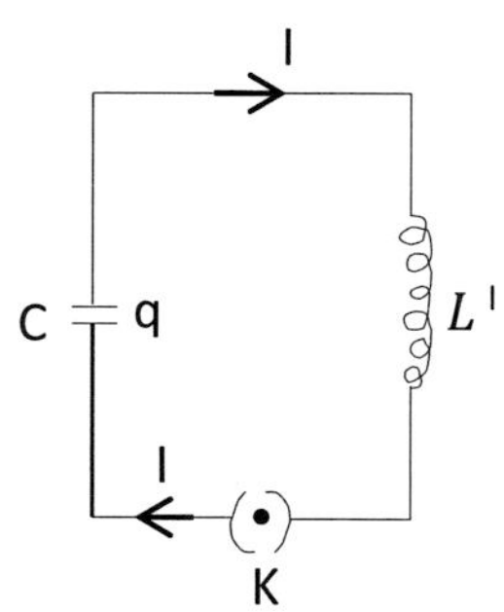

Fig. (3.14): *L – C Circuit*

Let 'q' be the charge on the Capacitor and I be the current flowing in the circuit at any instant of time 't'.
The energy in the Inductor is given by

$$U_B(Magnetic\ Energy) = \frac{1}{2}L^{\mathsf{I}}I^2 \text{ --------------- (1)}$$

This magnetic energy (U_B) is analogous to kinetic energy (T) in a mechanical system

$$U_B = \frac{1}{2}L^{\mathsf{I}}I^2 \quad ; \quad T = \frac{1}{2}mv^2$$

$$L^{\mathsf{I}} = Charge\ inertia \quad ; \quad m = Mass\ inertia$$

$$I = \frac{dq}{dt} \quad ; \quad V = \frac{dx}{dt}$$

Here Charge (q) is playing the role of displacement (x)

$$T = U_B = \frac{1}{2}L^{\mathsf{I}}I^2 \text{ ---------------------- (2)}$$

$$T = U_B = \frac{1}{2}L^{\mathsf{I}}\left(\frac{dq}{dt}\right)^2 \quad \left[We\ know\ that\ I = \frac{dq}{dt}\right]$$

$$T = U_B = \frac{1}{2}L^{\mathsf{I}}(\dot{q})^2 \text{ ------------------ (3)}$$

The electric potential energy of the circuit is the energy stored in electric field of capacitor.

$$V = U_E = \frac{1}{2}\frac{q^2}{C} = \frac{q^2}{2C} \text{ -------------- (4)}$$

Lagrangian of the system is given by

$$L = T - V \text{ --------------------------- (5)}$$

Now put the values of equations (3) & (4) in equation (5), we get

$$L = \frac{1}{2}L^{\mathsf{I}}(\dot{q})^2 - \frac{q^2}{2C} \text{ ---------------- (6)}$$

This equation (6) represents the Lagrangian of L – C circuit.
Lagrange's equation of motion is given by

$$\frac{d}{dt}\left(\frac{\partial L}{\partial \dot{q}}\right) - \frac{\partial L}{\partial q} = 0 \text{ ---------------- (7)}$$

Now partially differentiating equation (6) w. r. to $\dot{q}$, we get

$$\frac{\partial L}{\partial \dot{q}} = \frac{\partial}{\partial \dot{q}}\left\{\frac{1}{2}L^{\prime}(\dot{q})^2 - \frac{q^2}{2C}\right\}$$

$$\frac{\partial L}{\partial \dot{q}} = \frac{1}{2}L^{\prime}2\dot{q} - 0$$

$$\frac{\partial L}{\partial \dot{q}} = L^{\prime}\dot{q} \qquad \text{----------------- (8)}$$

Again partially differentiating equation (6) w. r. to q, we get

$$\frac{\partial L}{\partial q} = \frac{\partial}{\partial q}\left\{\frac{1}{2}L^{\prime}(\dot{q})^2 - \frac{q^2}{2C}\right\}$$

$$\frac{\partial L}{\partial q} = -\frac{q}{C} \qquad \text{---------------- (9)}$$

Put the values of equations (8) & (9) in equation (7) we get

$$\frac{d}{dt}(L^{\prime}\dot{q}) - \left(-\frac{q}{C}\right) = 0$$

$$L^{\prime}\ddot{q} + \frac{q}{C} = 0$$

Dividing on both sides by $L^{\prime}$ we get

$$\frac{L^{\prime}\ddot{q}}{L^{\prime}} + \frac{q}{CL^{\prime}} = \frac{0}{L^{\prime}}$$

$$\boldsymbol{\ddot{q} + \omega^2 q = 0} \qquad \text{----------------- (10)}$$

Where $\omega^2 = \frac{1}{CL^{\prime}}$

$$\omega = \sqrt{\frac{1}{CL^{\prime}}} = \frac{1}{\sqrt{CL^{\prime}}}$$

$$2\pi\vartheta = \frac{1}{\sqrt{CL^{\prime}}}$$

$$\vartheta = \frac{1}{2\pi\sqrt{CL^{\prime}}}$$

Equation (10) represents the required equation for L – C circuit.

3.14. INVARIANCE OF LAGRANGE'S EQUATION UNDER GALILEAN TRANSFORMATION

Let us consider 2 inertial frame of reference $S \,\&\, S^{\mathsf{I}}$ having Cartesian co – ordinates axis as $x, y, z \,\&\, x^{\mathsf{I}}, y^{\mathsf{I}}, z^{\mathsf{I}}$ and origin $O \,\&\, O^{\mathsf{I}}$ respectively. At time $t = o$, both the frames are at rest. So that their origin $O \,\&\, O^{\mathsf{I}}$ coincides with each other. Now let frame S^{I} starts moving with constant velocity v_0 along positive (+ve) direction of $x - axis$. Consider a moving particle at point P. The co – ordinates of point P w. r. to observer 'O' in frame S are (x, y, z, t) & w. r. to observer O^{I} in frame S^{I} are $(x^{\mathsf{I}}, y^{\mathsf{I}}, z^{\mathsf{I}}, t)$.

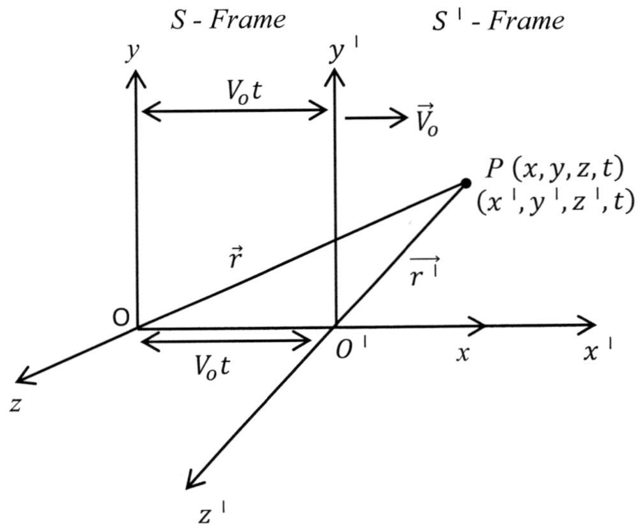

Fig. (3.15): Two Inertial Frames of Reference $S \,\&\, S^{\mathsf{I}}$

According to Galilean transformation equations, we have

$$x^{\mathsf{I}} = x - V_o t$$
$$y^{\mathsf{I}} = y$$
$$z^{\mathsf{I}} = z$$
$$t^{\mathsf{I}} = t$$

Let $\vec{r} \,\&\, \vec{r}^{\,\mathsf{I}}$ be the position vectors of the point P in frames $S \,\&\, S^{\mathsf{I}}$ respectively.

Using triangles law of vector addition in $\Delta OO^{\mathsf{I}}P$, we get

$$OO^{\mathsf{I}} + O^{\mathsf{I}}P = \vec{OP}$$

$$\vec{v_o} t + \vec{r}^{\,\mathsf{I}} = \vec{r}$$

$$\vec{r}^{\,|} = \vec{r} - \overrightarrow{v_o}t$$

Differentiating both sides w.r.to. 't' we get

$$\frac{d\vec{r}^{\,|}}{dt} = \frac{d\vec{r}}{dt} - \frac{d\,(\overrightarrow{v_o}t)}{dt}$$

$$\frac{d\vec{r}^{\,|}}{dt} = \frac{d\vec{r}}{dt} - \overrightarrow{v_o}\frac{dt}{dt}$$

$$\frac{d\vec{r}^{\,|}}{dt} = \frac{d\vec{r}}{dt} - \overrightarrow{v_o}$$

$$\vec{v}^{\,|} = \vec{v} - \overrightarrow{v_o} \qquad \text{---------------- (1)}$$

Where $\vec{v}^{\,|} = \frac{d\vec{r}^{\,|}}{dt}$ = Velocity of the particle at point 'P' in frame $S^{\,|}$

and $\vec{v} = \frac{d\vec{r}}{dt}$ = Velocity of the particle at point 'P' in frame S

Let L & $L^{\,|}$ be Lagrangian of the particle in frame S & $S^{\,|}$ respectively.

$$L = \frac{1}{2}mv^2 - V \qquad \text{--------------- (2)}$$

$$L^{\,|} = \frac{1}{2}mv^{\,|2} - V$$

$$L^{\,|} = \frac{1}{2}m.\vec{v}^{\,|}.\vec{v}^{\,|} - V \qquad \text{--------------- (3)}$$

Put the value of equation (1) in RHS of equation (3) we get

$$L^{\,|} = \frac{1}{2}m\,(\vec{v} - \overrightarrow{v_o})(\vec{v} - \overrightarrow{v_o}) - V$$

$$L^{\,|} = \frac{1}{2}m\,(\vec{v}.\vec{v} - \vec{v}.\overrightarrow{v_o} - \overrightarrow{v_o}.\vec{v} - \overrightarrow{v_o}.\overrightarrow{v_o}) - V$$

$$L^{\,|} = \frac{1}{2}m\,(v^2 - \vec{v}.\overrightarrow{v_o} - \overrightarrow{v_o}.\vec{v} - {v_o}^2) - V$$

$$L^{\,|} = \frac{1}{2}m\,(v^2 - 2\vec{v}.\overrightarrow{v_o} + {v_o}^2) - V$$

$$L^{\,|} = \frac{1}{2}mv^2 - \frac{1}{2}m(2\vec{v}.\overrightarrow{v_o}) + \frac{1}{2}m\,{v_o}^2 - V$$

$$L^{\,|} = \frac{1}{2}mv^2 - m\vec{v}.\overrightarrow{v_o} + \frac{1}{2}m\,{v_o}^2 - V$$

$$L^{\,|} = \left(\frac{1}{2}mv^2 - V\right) + \frac{1}{2}m\,{v_o}^2 - m\vec{v}.\overrightarrow{v_o} \qquad \text{-------------- (4)}$$

Put the value of equation (2) in equation (4) we get

$$L^{\prime} = L + \frac{1}{2}m\, v_o{}^2 - m\vec{v}.\overrightarrow{v_o}$$

$$L^{\prime} = L + \frac{1}{2}m\, v_o{}^2\frac{dt}{dt} - m\frac{d\vec{r}}{dt}.\overrightarrow{v_o} \qquad \begin{bmatrix} We\ know\ that \\ \vec{v} = \frac{d\vec{r}}{dt} \end{bmatrix}$$

$$L^{\prime} = L + \frac{d}{dt}\left(\frac{1}{2}m\, v_o{}^2 t\right) - \frac{d}{dt}(m\vec{r}\,\overrightarrow{v_o})$$

$$L^{\prime} = L + \frac{d}{dt}\left(\frac{1}{2}m\, v_o{}^2 t - m\vec{r}\,\overrightarrow{v_o}\right)$$

$$L^{\prime} = L + \frac{d}{dt}\, F\,(\vec{r}, t)$$

$$L^{\prime} = L + \frac{dF}{dt}$$

Where
$\vec{F} = \vec{F}(\vec{r}, t) = \frac{1}{2}m\, v_o{}^2 t - m\vec{r}\,\overrightarrow{v_o}$ = Gauge function for the Lagrangian

We know that Lagrange's equation is invariant under Gauge transformation

$$\frac{d}{dt}\left(\frac{\partial L^{\prime}}{\partial \dot{q}_k}\right) - \frac{\partial L^{\prime}}{\partial q_k} = 0$$

Thus, Lagrange's equation is invariant under Galilean transformation.

3.15. INVARIANCE OF LAGRANGE'S EQUATION UNDER GAUGE TRANSFORMATION

Let 'L' is the Lagrangian for a system of 'n' degrees of freedom, satisfying Lagrange's equation

i.e. $$\frac{d}{dt}\left(\frac{\partial L}{\partial \dot{q}_k}\right) - \frac{\partial L}{\partial q_k} = 0 \qquad \text{-------------------- (1)}$$

Then it can be shown that $L^{\prime} = L + \frac{dF}{dt}$ also satisfy the Lagrange's equation where F is an arbitrary function such as $F = F\,(q_1, q_2, \dots\dots\dots\, q_n, t)$ is called Gauge function for the Lagrangian.

Proof:

As $$F = F\,(q_1, q_2, \dots\dots\dots\, q_n, t) \qquad \text{----------------- (2)}$$

Differentiating both sides w. r. to 't' we get

$$\frac{dF}{dt} = \frac{\partial F}{\partial q_1}\frac{dq_1}{dt} + \frac{\partial F}{\partial q_2}\frac{dq_2}{dt} + ------ + \frac{\partial F}{\partial q_n}\frac{dq_n}{dt} + \frac{\partial F}{\partial t}\frac{dt}{dt}$$

$$\frac{dF}{dt} = \frac{\partial F}{\partial q_1}\dot{q}_1 + \frac{\partial F}{\partial q_2}\dot{q}_2 + ------ + \frac{\partial F}{\partial q_n}\dot{q}_n + \frac{\partial F}{\partial t}.1$$

$$\frac{dF}{dt} = \sum_{k=1}^{n}\frac{\partial F}{\partial q_k}\dot{q}_k + \frac{\partial F}{\partial t} \quad \text{----------------- (3)}$$

We have to prove $\frac{d}{dt}\left(\frac{\partial L'}{\partial \dot{q}_k}\right) - \frac{\partial L'}{\partial q_k} = 0$ ----------------- (4)

If we prove LHS of equation (4) to be equal to zero, then L' satisfy Lagrange's equation.

$$\text{LHS} = \frac{d}{dt}\left(\frac{\partial L'}{\partial \dot{q}_k}\right) - \frac{\partial L'}{\partial q_k}$$

$$\text{LHS} = \frac{d}{dt}\left[\frac{\partial}{\partial \dot{q}_k}\left(L + \frac{dF}{dt}\right)\right] - \frac{\partial}{\partial q_k}\left(L + \frac{dF}{dt}\right) \qquad \begin{bmatrix} We\ know\ that \\ L' = L + \frac{dF}{dt} \end{bmatrix}$$

$$\text{LHS} = \frac{d}{dt}\left[\frac{\partial L}{\partial \dot{q}_k} + \frac{\partial}{\partial \dot{q}_k}\left(\frac{dF}{dt}\right)\right] - \frac{\partial L}{\partial q_k} - \frac{\partial}{\partial q_k}\left(\frac{dF}{dt}\right)$$

$$\text{LHS} = \frac{d}{dt}\left(\frac{\partial L}{\partial \dot{q}_k}\right) + \frac{d}{dt}\left[\frac{\partial}{\partial \dot{q}_k}\left(\frac{dF}{dt}\right)\right] - \frac{\partial L}{\partial q_k} - \frac{\partial}{\partial q_k}\left(\frac{dF}{dt}\right)$$

$$\text{LHS} = \frac{d}{dt}\left(\frac{\partial L}{\partial \dot{q}_k}\right) - \frac{\partial L}{\partial q_k} + \frac{d}{dt}\left[\frac{\partial}{\partial \dot{q}_k}\left(\frac{dF}{dt}\right)\right] - \frac{\partial}{\partial q_k}\left(\frac{dF}{dt}\right)$$

$$\text{LHS} = 0 + \frac{d}{dt}\left[\frac{\partial}{\partial \dot{q}_k}\left(\frac{dF}{dt}\right)\right] - \frac{\partial}{\partial q_k}\left(\frac{dF}{dt}\right) \qquad \begin{bmatrix} We\ know\ that \\ \frac{d}{dt}\left(\frac{\partial L}{\partial \dot{q}_k}\right) - \frac{\partial L}{\partial q_k} = 0 \end{bmatrix}$$

$$\text{LHS} = \frac{d}{dt}\left[\frac{\partial}{\partial \dot{q}_k}\left(\frac{dF}{dt}\right)\right] - \frac{\partial}{\partial q_k}\left(\frac{dF}{dt}\right) \quad \text{---------------- (5)}$$

Put the value of equation (3) in equation (5) we get

$$\text{LHS} = \frac{d}{dt}\left[\frac{\partial}{\partial \dot{q}_k}\left(\sum_{k=1}^{n}\frac{\partial F}{\partial q_k}\dot{q}_k + \frac{\partial F}{\partial t}\right)\right] - \frac{\partial}{\partial q_k}\left(\sum_{k=1}^{n}\frac{\partial F}{\partial q_k}\dot{q}_k + \frac{\partial F}{\partial t}\right)$$

$$\text{LHS} = \frac{d}{dt}\left[\sum_{k=1}^{n}\frac{\partial}{\partial \dot{q}_k}\left(\frac{\partial F}{\partial q_k}\dot{q}_k\right) + \frac{\partial}{\partial \dot{q}_k}\left(\frac{\partial F}{\partial t}\right)\right] - \sum_{k=1}^{n}\frac{\partial}{\partial q_k}\left(\frac{\partial F}{\partial q_k}\dot{q}_k\right) - \frac{\partial}{\partial q_k}\left(\frac{\partial F}{\partial t}\right)$$

LHS = $\frac{d}{dt}\left\{\sum_{k=1}^{n}\left[\frac{\partial F}{\partial q_k}\frac{\partial \dot{q}_k}{\partial \dot{q}_k} + \dot{q}_k\frac{\partial}{\partial \dot{q}_k}\left(\frac{\partial F}{\partial q_k}\right)\right] + \frac{\partial}{\partial t}\left(\frac{\partial F}{\partial \dot{q}_k}\right)\right\} - \sum_{k=1}^{n}\left[\frac{\partial F}{\partial q_k}\frac{\partial \dot{q}_k}{\partial q_k} + \dot{q}_k\frac{\partial}{\partial q_k}\left(\frac{\partial F}{\partial q_k}\right)\right] - \frac{\partial}{\partial q_k}\left(\frac{\partial F}{\partial t}\right)$

LHS = $\frac{d}{dt}\left\{\sum_{k=1}^{n}\left[\frac{\partial F}{\partial q_k}\frac{\partial \dot{q}_k}{\partial \dot{q}_k} + \dot{q}_k\frac{\partial}{\partial q_k}\left(\frac{\partial F}{\partial \dot{q}_k}\right)\right] + \frac{\partial}{\partial t}\left(\frac{\partial F}{\partial \dot{q}_k}\right)\right\} - \sum_{k=1}^{n}\left[\frac{\partial F}{\partial q_k}\frac{\partial \dot{q}_k}{\partial q_k} + \dot{q}_k\frac{\partial}{\partial q_k}\left(\frac{\partial F}{\partial q_k}\right)\right] - \frac{\partial}{\partial q_k}\left(\frac{\partial F}{\partial t}\right)$ -------------------- (6)

Here F is not a function of $\dot{q}_k$ so we have

$\frac{\partial F}{\partial \dot{q}_k} = 0$, $\frac{\partial \dot{q}_k}{\partial \dot{q}_k} = 0$, $\frac{\partial \dot{q}_k}{\partial q_k} = 0$ ---------------- (7) $\begin{bmatrix} We\ have\ from\ equation\ (2) \\ F = F\,(q_1, q_2, \dots\dots\dots\ q_n, t) \end{bmatrix}$

Put the values of equation (7) in equation (6) we have

LHS = $\frac{d}{dt}\left\{\sum_{k=1}^{n}\left[\frac{\partial F}{\partial q_k}(1) + \dot{q}_k\frac{\partial}{\partial q_k}(0)\right] + \frac{\partial}{\partial t}(0)\right\} - \sum_{k=1}^{n}\left[\frac{\partial F}{\partial q_k}(0) + \dot{q}_k\frac{\partial}{\partial q_k}\left(\frac{\partial F}{\partial q_k}\right)\right] - \frac{\partial}{\partial q_k}\left(\frac{\partial F}{\partial t}\right)$

LHS = $\frac{d}{dt}\sum_{k=1}^{n}\left(\frac{\partial F}{\partial q_k}\right) - \sum_{k=1}^{n}\dot{q}_k\frac{\partial}{\partial q_k}\left(\frac{\partial F}{\partial q_k}\right) - \frac{\partial}{\partial q_k}\left(\frac{\partial F}{\partial t}\right)$

LHS = $\sum_{k=1}^{n}\frac{\partial}{\partial q_k}\left(\frac{dF}{dt}\right) - \sum_{k=1}^{n}\dot{q}_k\frac{\partial}{\partial q_k}\left(\frac{\partial F}{\partial q_k}\right) - \frac{\partial}{\partial q_k}\left(\frac{\partial F}{\partial t}\right)$ -------- (8)

Put the value of equation (3) equation (8) we have

LHS = $\sum_{k=1}^{n}\frac{\partial}{\partial q_k}\left(\sum_{k=1}^{n}\frac{\partial F}{\partial q_k}\dot{q}_k + \frac{\partial F}{\partial t}\right) - \sum_{k=1}^{n}\dot{q}_k\frac{\partial}{\partial q_k}\left(\frac{\partial F}{\partial q_k}\right) - \frac{\partial}{\partial q_k}\left(\frac{\partial F}{\partial t}\right)$

LHS = $\sum_{k=1}^{n}\frac{\partial}{\partial q_k}\left(\frac{\partial F}{\partial q_k}.\dot{q}_k\right) + \frac{\partial}{\partial q_k}\left(\frac{\partial F}{\partial t}\right) - \sum_{k=1}^{n}\dot{q}_k\frac{\partial}{\partial q_k}\left(\frac{\partial F}{\partial q_k}\right) - \frac{\partial}{\partial q_k}\left(\frac{\partial F}{\partial t}\right)$

LHS = $\sum_{k=1}^{n}\dot{q}_k\frac{\partial}{\partial q_k}\left(\frac{\partial F}{\partial q_k}\right) + \frac{\partial}{\partial q_k}\left(\frac{\partial F}{\partial t}\right) - \sum_{k=1}^{n}\dot{q}_k\frac{\partial}{\partial q_k}\left(\frac{\partial F}{\partial q_k}\right) - \frac{\partial}{\partial q_k}\left(\frac{\partial F}{\partial t}\right)$

LHS = 0

Thus $\frac{d}{dt}\left(\frac{\partial L'}{\partial \dot{q}_k}\right) - \frac{\partial L'}{\partial q_k} = 0$, Hence L' satisfy the Lagrange's equation.

3.16. LAGRANGIAN FORMULATION OF CONSERVATION THEOREMS

For the purpose let u first define generalized momentum and cyclic co – ordinates.

Generalized Momentum

Consider a conservative system, i.e., a system in which the forces are derivable from a potential function V dependent on position only; then

$$\frac{\partial L}{\partial \dot{q}_k} = \frac{\partial (T-V)}{\partial \dot{q}_k} = \frac{\partial T}{\partial \dot{q}_k} - \frac{\partial V}{\partial \dot{q}_k}$$

$= \frac{\partial T}{\partial \dot{q}_k}$ because potential V depends only on position or $\frac{\partial V}{\partial \dot{q}_k} = 0$

$$= \frac{\partial}{\partial \dot{q}_k} \sum_k \frac{1}{2} m_k \dot{q}_k^{\,2}$$

$$= m_k \dot{q}_k = p_k$$

p_k is called the *generalized momentum* associated with generalized co – ordinate q_k

$$p_k = \frac{\partial L}{\partial \dot{q}_k} \qquad \text{-------------------- (1)}$$

Thus, the generalized momentum corresponding to the generalized co – ordinate q_k is defined as the quantity $\frac{\partial L}{\partial \dot{q}_k}$ and is represented by p_k, i.e.,

Cyclic or Ignorable co – ordinates

If the Lagrangian of a system is not the function of a given co – ordinate q_k is said to be cyclic (or Ignorable).

In such case $\quad \frac{\partial L}{\partial q_k} = 0 \qquad$ ------------------- (2)

(a) Conservation theorem for generalized momentum:

According to Lagrange's equation

$$\frac{d}{dt}\left(\frac{\partial L}{\partial \dot{q}_k}\right) - \frac{\partial L}{\partial q_k} = 0 \qquad \text{---------------- (3)}$$

If the generalized co – ordinate q_k is cyclic (or ignorable), then

We have $\quad \frac{\partial L}{\partial q_k} = 0 \qquad [From\ equation\ (2)]$

With this substitution (3) becomes

$$\frac{d}{dt}\left(\frac{\partial L}{\partial \dot{q}_k}\right) = 0$$

Or $$\frac{\partial L}{\partial \dot{q}_k} = Constant$$

i.e., $$p_k = Constant \quad [From\ equation\ (1)]$$

This gives the conservation theorem for generalized momentum which states that the *generalized momentum corresponding to a cyclic (or ignorable) co – ordinate is conserved.*

(b) Conservation theorem for energy:

Consider a conservative system, i.e., a system in which the forces re derivable from a potential function V dependent on position only. The Lagrangian L is given by

$$L = L(q_1, q_2, - - - q_k - - - \dot{q}_1, \dot{q}_2, - - - \dot{q}_k, - - - t)$$

Let us assume that L does not depend upon time explicitly.
Then
The total time derivative of L is given by

$$\frac{dL}{dt} = \frac{\partial L}{\partial q_1}\dot{q}_1 + \frac{\partial L}{\partial q_2}\dot{q}_2 + - - - + \frac{\partial L}{\partial q_k}\dot{q}_k + - - - + \frac{\partial L}{\partial \dot{q}_1}\ddot{q}_1 + \frac{\partial L}{\partial \dot{q}_2}\ddot{q}_2 + - - - + \frac{\partial L}{\partial \dot{q}_k}\ddot{q}_k + - - -$$

$$\frac{dL}{dt} = \sum_k \frac{\partial L}{\partial q_k}\dot{q}_k + \sum_k \frac{\partial L}{\partial \dot{q}_k}\ddot{q}_k \qquad \text{------------------ (4)}$$

Lagrange's equation is given by

$$\frac{d}{dt}\left(\frac{\partial L}{\partial \dot{q}_k}\right) - \frac{\partial L}{\partial q_k} = 0$$

Or $$\frac{d}{dt}\left(\frac{\partial L}{\partial \dot{q}_k}\right) = \frac{\partial L}{\partial q_k}$$

With this substitution equation (4) becomes

$$\frac{dL}{dt} = \sum_k \frac{d}{dt}\left(\frac{\partial L}{\partial \dot{q}_k}\right)\dot{q}_k + \sum_k \frac{\partial L}{\partial \dot{q}_k}\ddot{q}_k$$

$$\frac{dL}{dt} = \sum_k \frac{d}{dt}\left(\dot{q}_k \frac{\partial L}{\partial \dot{q}_k}\right)$$

Or $$\frac{d}{dt}\left(\sum \dot{q}_k \frac{\partial L}{\partial \dot{q}_k} - L\right) = 0$$

Thereby indicating that $\sum \dot{q}_k \frac{\partial L}{\partial \dot{q}_k} - L$ = Constant -------------- (5)

But $$\frac{\partial L}{\partial \dot{q}_k} = \frac{\partial}{\partial \dot{q}_k}(T - V) = \frac{\partial T}{\partial \dot{q}_k}$$

$$\frac{\partial L}{\partial \dot{q}_k} = \frac{\partial}{\partial \dot{q}_k}\left(\sum \frac{1}{2} m_k \dot{q}_k{}^2\right)$$

Because $T = \sum \frac{1}{2} m_k \dot{q}_k{}^2$, Where m_k is the mass of kth particle.

∴ $$\frac{\partial L}{\partial \dot{q}_k} = m_k \dot{q}_k$$

Substituting the values of $\frac{\partial L}{\partial \dot{q}_k}$ from above equation in (5), we get

$$\sum m_k \dot{q}_k{}^2 - L = Constant$$

Or $$2T - L = Constant$$

Or $$2T - (T - V) = Constant$$

Or $$(T + V) = Constant$$

∴ Total mechanical energy $T + V = Total\ Energy = Constant$

Thus, the energy conservation theorem states that, ***if the Lagrangian function does not contain the time explicitly, the total energy of the conservative system is conserved.***

(c) Conservation Theorem for Linear Momentum:

Consider a conservative system so that the potential energy V is dependent on position only and the kinetic energy is independent of position, i.e.,

$$\frac{\partial V}{\partial \dot{q}_k} = 0 \text{ and } \frac{\partial T}{\partial q_k} = 0 \qquad \text{-------------- (6)}$$

The Lagrange's equation for such a system may written as

$$\frac{d}{dt}\left(\frac{\partial L}{\partial \dot{q}_k}\right) - \frac{\partial L}{\partial q_k} = 0$$

i.e., $\frac{d}{dt}\left(\frac{\partial L}{\partial \dot{q}_k}\right) + \frac{\partial V}{\partial q_k} = 0$ [*Using equation* (6)]

So $\dot{p}_k = \frac{d}{dt}\left(\frac{\partial L}{\partial \dot{q}_k}\right) = -\frac{\partial V}{\partial q_k} = Q_k$ -------------- (7)

Now if we show that the generalized force Q_k represents the component of total force along the direction of translation of the system and generalized momentum p_k is the component of total linear momentum along the same direction, then equation (7) will represent the equation of motion for total linear momentum.

The generalized force Q_k is given by

$$Q_k = \sum_i F_i \cdot \frac{\partial r_i}{\partial q_k}$$

If $\hat{n}$ is the unit vector along the direction of translation of the system, then

$$\delta r_i = \hat{n}\delta q_k$$

i.e., $\frac{\partial r_i}{\partial q_k} = \hat{n}$

So $Q_k = \sum_i F_i \cdot \hat{n} = \hat{n} \cdot F$ ----------------- (8)

Which represents the component of total force along the direction on $\hat{n}$ $(i.e., translation)$

The kinetic energy, $T = \frac{1}{2}\sum_i m_i \dot{r}_i^2$

So, the generalized momentum,

$$p_k = \frac{\partial L}{\partial \dot{q}_k} = \frac{\partial T}{\partial \dot{q}_k} = \sum_i m_i \dot{r}_i \cdot \frac{\partial \dot{r}_i}{\partial \dot{q}_k}$$

$$= \sum_i m_i v_i \cdot \frac{\partial r_i}{\partial q_k} \quad \text{------------ (9)}$$

$$= \sum_i m_i v_i \cdot \hat{n}$$

Substituting values of p_k *and* Q_k in equation (7), we get

$$\frac{d}{dt}\left(\sum m_i v_i . \hat{n}\right) = \hat{n}.F$$

Thus, the equation of motion for total linear momentum.

$$\frac{dP}{dt} = F$$

If $\quad F = 0, \quad P = Constant$

This gives the conservation theorem for linear momentum and states that ***if a given component of the total applied force vanishes, the corresponding component of the linear momentum is conserved.***

(d) Conservation Theorem for Angular Momentum:

Consider a conservative system so that the potential energy V is dependent of position only and the kinetic energy is independent of position.

$$\frac{\partial V}{\partial \dot{q}_k} = 0 \text{ and } \frac{\partial T}{\partial q_k} = 0 \qquad \text{------------- (10)}$$

The Lagrange's equation for such a system is

$$\frac{d}{dt}\left(\frac{\partial L}{\partial \dot{q}_k}\right) - \frac{\partial L}{\partial q_k} = 0$$

$$\frac{d}{dt}\left(\frac{\partial L}{\partial \dot{q}_k}\right) = \frac{\partial L}{\partial q_k} = \frac{\partial V}{\partial q_k}$$

$$\dot{p}_k = Q_k \qquad \text{-------------- (11)}$$

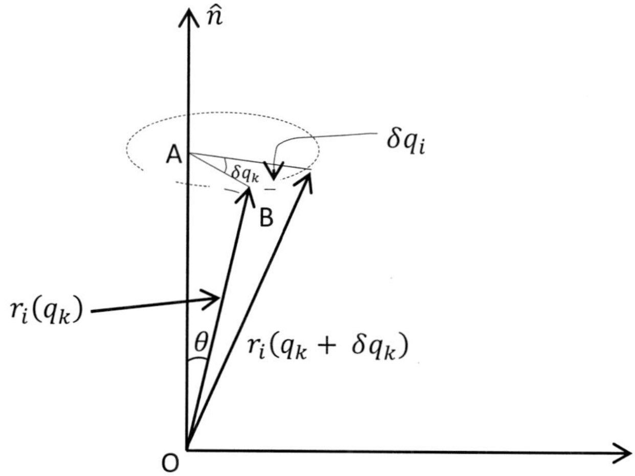

Fig. (3.16)

Now if we show that with rotation co – ordinate q_k, the generalized force Q_k is the component of total applied torque about the axis of rotation and the generalized momentum along the same axis, then equation (11) will represent the equation of motion for the total angular momentum.

The generalized force Q_k is given by

$$Q_k = \sum_i F_i . \frac{\partial r_i}{\partial q_k}$$

The magnitude of change in position co – ordinate $r_i(q_k)$ due to a change in rotation co – ordinate q_k is given by

$$|\delta r_i| = AB\delta q_k$$

$$= r_i Sin\theta \delta q_k \qquad \text{---------------} (12)$$

$$\left|\frac{\partial r_i}{\partial q_k}\right| = r_i Sin\theta$$

If $\hat{n}$ is the unit vector along the axis of rotation, equation (12) may be written as

$$\frac{\partial r_i}{\partial q_k} = \hat{n} \, X \, r_i \qquad \text{---------------} (13)$$

Now the generalized force can be expressed as

$$Q_k = \sum_i F_i . \hat{n} \, X \, r_i$$

$$Q_k = \sum_i \hat{n} \, . (r_i X \, F_i)$$

$$Q_k = \sum_i \hat{n} \, . \vec{\tau}_j$$

$$Q_k = \hat{n} \, . \sum_i \vec{\tau}_j$$

$$Q_k = \hat{n} \, . \vec{\tau}_j \qquad \text{---------------} (14)$$

Where $\tau = \sum \tau_j$ is the total torque. Equation (14) shows that Q_k is the component of the total torque along the axis of rotation.

Also $\quad p_k = \frac{\partial L}{\partial \dot{q}_k} = \frac{\partial T}{\partial \dot{q}_k} = \sum_i m_i \dot{r}_i . \frac{\partial \dot{r}_i}{\partial \dot{q}_k}$

$$= \sum_i m_i v_i . \frac{\partial r_i}{\partial q_k} \quad [Using\ equation\ (9)]$$

$$= \sum_i m_i v_i . \hat{n}\, X\, r_i \quad [Using\ equation\ (13)]$$

$$= \sum_i \hat{n} . r_i\, X\, m_i V_i$$

$$= \hat{\boldsymbol{n}} . \sum \dot{\boldsymbol{L}}_i$$

$$= \hat{\boldsymbol{n}} . \boldsymbol{L} \qquad \text{---------------} \ (15)$$

Where $L = \sum_i L_i$ is the total angular momentum i.e.,

Substituting Q_k and p_k from (14), we get

$$\frac{d}{dt}(\hat{n} . L) = \hat{n} . \vec{\tau}$$

This equation represents the equation of motion for the total angular momentum of the system.

If $\hat{n} . \vec{\tau} = 0;$ then $\hat{n} . L = Constant$

Thus the conservation theorem for angular momentum states ***if the component of applied torque along the axis of rotation vanishes, then the total angular momentum along the axis of rotation is conserved.***

3.17. NEWTON'S EQUATION OF MOTION FROM LAGRANGE'S EQUATION (Newton's 2nd Law Using Lagrange's Equation)

Consider the motion of a particle of mass 'm' using Cartesian co – ordinates as generalized co – ordinates; deduce Newton's equation of motion from Lagrangian equation.

Solution:

The generalized form of Lagrange's equation is given by

$$\frac{d}{dt}\left(\frac{\partial T}{\partial \dot{q}_k}\right) - \frac{\partial T}{\partial q_k} = G_k \qquad \text{----------------} \ (1)$$

Where

T = Kinetic energy of particle

G_k = Generalized force

q_k = Generalized Co – ordinate

Where K = 1, 2, 3 …………

Here generalized co – ordinates are Cartesian co – ordinates,
So, we have $q_1 = x, \quad q_2 = y, \quad q_3 = z$

Generalized force components are
$G_1 = F_x\,, \quad G_2 = F_y\,, \quad G_3 = F_z\,,$

For x – coordinate equation (1) takes the form as

$$\frac{d}{dt}\left(\frac{\partial T}{\partial \dot{x}}\right) - \frac{\partial T}{\partial x} = F_x \qquad \text{------------------ (2)}$$

The Kinetic Energy 'T' of the system is given by

$$T = \frac{1}{2}\, mV^2 \qquad \text{------------------ (3)}$$

Now $\quad \vec{V} = V_x\hat{\imath} + V_y\hat{\jmath} + V_z\hat{k}$

$$V = \sqrt{V_x{}^2 + V_y{}^2 + V_z{}^2}$$

$$V^2 = V_x{}^2 + V_y{}^2 + V_z{}^2$$

$$V^2 = \left(\frac{dx}{dt}\right)^2 + \left(\frac{dy}{dt}\right)^2 + \left(\frac{dz}{dt}\right)^2$$

$$V^2 = \dot{x}^2 + \dot{y}^2 + \dot{z}^2 \qquad \text{------------------ (4)}$$

Put the value of equation (4) in equation (3), we get

$$T = \frac{1}{2}\, m\,(\dot{x}^2 + \dot{y}^2 + \dot{z}^2)^2 \qquad \text{------------------ (5)}$$

Partially differentiating equation (5) on both sides w. r. to $\dot{x}$

$$\frac{\partial T}{\partial \dot{x}} = \frac{1}{2}\, m\,(2\dot{x} + 0 + 0)$$

$$\frac{\partial T}{\partial \dot{x}} = m\dot{x} \qquad \text{----------------- (6)}$$

Partially differentiating equation (5) on both sides w. r. to x

$$\frac{\partial T}{\partial x} = 0 \qquad \text{----------------- (7)}$$

Put the values of equation (6) & (7) in equation (2) we get

$$\frac{d}{dt}(m\dot{x}) - 0 = F_x$$

$$\frac{d}{dt}\left(m\frac{dx}{dt}\right) = F_x$$

$$\frac{d}{dt}(mV_x) = F_x \qquad \left[We\ know\ \frac{dx}{dt} = V_x\right]$$

$$\frac{d}{dt}P_x = F_x \qquad [We\ know\ mV_x = P_x]$$

$\therefore$ $$F_x = \frac{dP_x}{dt}$$

Similarly we can write

$$F_y = \frac{dP_y}{dt}$$

$$F_z = \frac{dP_z}{dt}$$

So, In general

$$\vec{\boldsymbol{F}} = \frac{d\vec{\boldsymbol{P}}}{dt}$$

This is the required Newton's equation of motion

3.18. LAGRANGIAN OF A CHARGED PARTICLE IN AN ELECTROMAGNETIC FIELD

The force acting on a charge 'q' moving with velocity $\vec{V}$ in an electric field $\vec{B}$ is given by

$$\vec{F} = q\vec{E} + q(\vec{V}\, X\, \vec{B})$$

$$\vec{F} = q[\vec{E} + (\vec{V}\, X\, \vec{B})] \text{ ----------- (1)}(Lorentz\ Force)$$

As $$\vec{B} = \nabla\, X\, \vec{A} \text{ ----------- (2)}(\vec{A}\ is\ magnetic\ vector\ potential)$$

From Maxwell's field equation we have

$$\nabla\, X\, \vec{E} + \frac{\partial B}{\partial t} = 0 \qquad \text{--------------- (3)}$$

Put the value of equation (3) in equation (2) we get

$$\nabla X \vec{E} + \frac{\partial}{\partial t}(\nabla X \vec{A}) = 0$$

$$\nabla X \left(\vec{E} + \frac{\partial \vec{A}}{\partial t}\right) = 0 \qquad \text{---------------} (4)$$

From vector calculas we know that curl of gradient of some scalar function '∅' is always zero. i.e.,

$$\nabla X (-\nabla \emptyset) = 0 \qquad \text{---------------} (5)$$

Comparing equation (4) & (5) we get

$$\vec{E} + \frac{\partial \vec{A}}{\partial t} = -\nabla \emptyset$$

$$\vec{E} = -\nabla \emptyset - \frac{\partial \vec{A}}{\partial t} \qquad \text{---------------} (6)$$

Put the values of equations (2) & (6) in equation (1) we get

$$\vec{F} = q\left[-\nabla \emptyset - \frac{\partial \vec{A}}{\partial t} + (\vec{V} X \nabla X \vec{A})\right]$$

Taking x component of $\vec{F}$ only we get

$$F_x = q\left[-\frac{\partial \emptyset}{\partial x} - \frac{\partial A_x}{\partial t} + \left(\vec{V} X \nabla X \vec{A}\right)_x\right] \qquad \text{------------} (7)$$

Let us find $\left(\vec{V} X \nabla X \vec{A}\right)_x$

$$\nabla X \vec{A} = \begin{vmatrix} \hat{\imath} & \hat{\jmath} & \hat{k} \\ \frac{\partial}{\partial x} & \frac{\partial}{\partial y} & \frac{\partial}{\partial z} \\ A_x & A_y & A_z \end{vmatrix} \qquad \begin{pmatrix} We\ know\ that \\ \nabla = \hat{\imath}\frac{\partial}{\partial x} + \hat{\jmath}\frac{\partial}{\partial y} + \hat{k}\frac{\partial}{\partial z} \\ \vec{A} = A_x\hat{\imath} + A_y\hat{\jmath} + A_z\hat{k} \end{pmatrix}$$

$$\nabla X \vec{A} = \hat{\imath}\left(\frac{\partial A_z}{\partial y} - \frac{\partial A_y}{\partial z}\right) - \hat{\jmath}\left(\frac{\partial A_z}{\partial x} - \frac{\partial A_x}{\partial z}\right) + \hat{k}\left(\frac{\partial A_y}{\partial x} - \frac{\partial A_x}{\partial y}\right)$$

$$\nabla X \vec{A} = \hat{\imath}\left(\frac{\partial A_z}{\partial y} - \frac{\partial A_y}{\partial z}\right) + \hat{\jmath}\left(-\frac{\partial A_z}{\partial x} + \frac{\partial A_x}{\partial z}\right) + \hat{k}\left(\frac{\partial A_y}{\partial x} - \frac{\partial A_x}{\partial y}\right)$$

$$\vec{V} X(\nabla X \vec{A}) = \begin{vmatrix} \hat{\imath} & \hat{\jmath} & \hat{k} \\ V_x & V_y & V_z \\ \left(\frac{\partial A_z}{\partial y} - \frac{\partial A_y}{\partial z}\right) & \left(-\frac{\partial A_z}{\partial x} + \frac{\partial A_x}{\partial z}\right) & \left(\frac{\partial A_y}{\partial x} - \frac{\partial A_x}{\partial y}\right) \end{vmatrix}$$

$$\left[\vec{V}\,X(\nabla\,X\,\vec{A})\right]_x = V_y\left(\frac{\partial A_y}{\partial x} - \frac{\partial A_x}{\partial y}\right) - V_z\left(-\frac{\partial A_z}{\partial x} + \frac{\partial A_x}{\partial z}\right)$$

$$\left[\vec{V}\,X(\nabla\,X\,\vec{A})\right]_x = V_y\frac{\partial A_y}{\partial x} - V_y\frac{\partial A_x}{\partial y} + V_z\frac{\partial A_z}{\partial x} - V_z\frac{\partial A_x}{\partial z}$$

Adding and subtracting $V_x\frac{\partial A_x}{\partial x}$ on RHS we get

$$\left[\vec{V}\,X(\nabla\,X\,\vec{A})\right]_x = V_x\frac{\partial A_x}{\partial x} + V_y\frac{\partial A_y}{\partial x} + V_z\frac{\partial A_z}{\partial x} - V_x\frac{\partial A_x}{\partial x} - V_y\frac{\partial A_x}{\partial y} - V_z\frac{\partial A_x}{\partial z}$$

$$\left[\vec{V}\,X(\nabla\,X\,\vec{A})\right]_x = V_x\frac{\partial A_x}{\partial x} + V_y\frac{\partial A_y}{\partial x} + V_z\frac{\partial A_z}{\partial x} - \left(V_x\frac{\partial A_x}{\partial x} + V_y\frac{\partial A_x}{\partial y} + V_z\frac{\partial A_x}{\partial z}\right) \text{------- (8)}$$

Now $\frac{\partial}{\partial x}(\vec{V}.\vec{A}) = \frac{\partial}{\partial x}(V_xA_x + V_yA_y + V_zA_z)$

$$\frac{\partial}{\partial x}(\vec{V}.\vec{A}) = V_x\frac{\partial A_x}{\partial x} + V_y\frac{\partial A_y}{\partial x} + V_z\frac{\partial A_z}{\partial x} \qquad \text{------------- (9)}$$

Now $A_x = A_x(x, y, z, t)$

So $dA_x = \frac{\partial A_x}{\partial x}dx + \frac{\partial A_x}{\partial y}dy + \frac{\partial A_x}{\partial z}dz + \frac{\partial A_x}{\partial t}dt$

Divide with 'dt' on both sides we get

$$\frac{dA_x}{dt} = \frac{\partial A_x}{\partial x}\frac{dx}{dt} + \frac{\partial A_x}{\partial y}\frac{dy}{dt} + \frac{\partial A_x}{\partial z}\frac{dz}{dt} + \frac{\partial A_x}{\partial t}\frac{dt}{dt}$$

$$\frac{dA_x}{dt} = \frac{\partial A_x}{\partial x}\frac{dx}{dt} + \frac{\partial A_x}{\partial y}\frac{dy}{dt} + \frac{\partial A_x}{\partial z}\frac{dz}{dt} + \frac{\partial A_x}{\partial t}$$

$$\frac{dA_x}{dt} - \frac{\partial A_x}{\partial t} = \frac{\partial A_x}{\partial x}V_x + \frac{\partial A_x}{\partial y}V_y + \frac{\partial A_x}{\partial z}V_z \qquad \text{------------- (10)}$$

Put the value of equation (9) & (10) in equation (8) we get

$$\left[\vec{V}\,X(\nabla\,X\,\vec{A})\right]_x = \frac{\partial}{\partial x}(\vec{V}.\vec{A}) - \left(\frac{dA_x}{dt} - \frac{\partial A_x}{\partial t}\right)$$

$$\left[\vec{V}\,X(\nabla\,X\,\vec{A})\right]_x = \frac{\partial}{\partial x}(\vec{V}.\vec{A}) - \frac{dA_x}{dt} + \frac{\partial A_x}{\partial t} \qquad \text{------------- (11)}$$

Put the value of equation (11) in equation (7) we get

$$F_x = q\left[-\frac{\partial \emptyset}{\partial x} - \frac{\partial A_x}{\partial t} + \frac{\partial}{\partial x}(\vec{V}.\vec{A}) - \frac{dA_x}{dt} + \frac{\partial A_x}{\partial t}\right]$$

$$F_x = q\left[-\frac{\partial \emptyset}{\partial x} + \frac{\partial}{\partial x}(\vec{V}.\vec{A}) - \frac{dA_x}{dt}\right]$$

$$F_x = q\left[-\frac{\partial}{\partial x}(\emptyset - \vec{V}.\vec{A}) - \frac{dA_x}{dt}\right] \quad \text{-------------} (12)$$

Now $\frac{\partial}{\partial V_x}(\vec{V}.\vec{A}) = \frac{\partial}{\partial V_x}(V_xA_x + V_yA_y + V_zA_z)$

$$\frac{\partial}{\partial V_x}(\vec{V}.\vec{A}) = A_x$$

$$\frac{\partial}{\partial V_x}(-\vec{V}.\vec{A}) = -A_x$$

$$\frac{\partial}{\partial V_x}(\emptyset - \vec{V}.\vec{A}) = -A_x$$

$$\frac{d}{dt}\left[\frac{\partial}{\partial V_x}(\emptyset - \vec{V}.\vec{A})\right] = -\frac{d}{dt}A_x \quad \text{--------------} (13)$$

Put the value of equation (13) in equation (12) we get

$$F_x = q\left[-\frac{\partial}{\partial x}(\emptyset - \vec{V}.\vec{A}) + \left\{\frac{d}{dt}\left[\frac{\partial}{\partial V_x}(\emptyset - \vec{V}.\vec{A})\right]\right\}\right]$$

$$F_x = -\frac{\partial}{\partial x}q(\emptyset - \vec{V}.\vec{A}) + \frac{d}{dt}\left[\frac{\partial}{\partial V_x}q(\emptyset - \vec{V}.\vec{A})\right] \quad \text{-----------} (14)$$

Comparing (14) with $F_x = -\frac{\partial U}{\partial x} + \frac{d}{dt}\frac{\partial U}{\partial V_x}$ we get
$U = q\frac{d}{dt}\left[\frac{\partial}{\partial V_x}(\emptyset - \vec{V}.\vec{A})\right]$ is called velocity dependent generalized potential

Now $L = T - U$

$$\boldsymbol{L = T - q(\emptyset - \vec{V}.\vec{A})}$$

This is the required Lagrangian of a Charged particle in an Electromagnetic field.

3.19. LAGRANGE'S EQUATIONS FOR ELECTRICAL CIRCUITS

The Lagrange's equations for electrical circuits containing finite number of inductances, capacitors and resistances may be written in the following form:

$$\frac{d}{dt}\left(\frac{\partial L_{El}}{\partial \dot{q}_k}\right) - \frac{\partial L_{El}}{\partial q_k} = Q_k \qquad \text{-------------- (1)}$$

Where L_{El} is the Lagrangian for electrical circuit and is given by

$$L_{El} = T_M - V_{El} \qquad \text{-------------- (2)}$$

Where

T_M is the ***magnetic energy*** of the electrical circuit and is analogous to the kinetic energy of the mechanical system.

V_{El} is the ***electrical energy*** of the electrical circuit and is analogous to the Potential energy of the mechanical system.

Q_k is the ***generalized force*** due to friction. The conservative forces are included in the Lagrangian L_{El}.

If the system does not contain frictional or dissipative forces, i.e., if $Q_k = 0$, equation (1) can be written as

$$\frac{d}{dt}\left(\frac{\partial L_{El}}{\partial \dot{q}_k}\right) - \frac{\partial L_{El}}{\partial q_k} = 0 \qquad \text{---------------- (3)}$$

3.20. NEWTONIAN AND LAGRANGIAN FORMALISMS

The Lagrangian formalism is not the result of a new theory but it is derivable from Newton's second law. In the Newtonian formalism, all the forces acting on the system, both active and internal forces must be taken into account. That is, the dynamical conditions must be known. But the Lagrangian method concentrates solely on active forces, completely ignoring the forces of constraints by formulating the problem in terms of generalized co – ordinates. This gives the advantage of selecting any suitable quantity such as linear momentum, angular momentum, velocity or angle as the generalized co – ordinates depending on the problem. Secondly, the Newtonian force momentum approach is vectorial in nature whereas the work – energy approach of Lagrangian method involves only scalar functions. In the Lagrangian approach, all the details are contained in a single scalar function, the Lagrangian of the system. Though the directional properties of the vectors are more helpful when dealing with simple systems, the formulation becomes difficult when the system becomes more complex.

The Lagrangian method is applicable to conservative forces only, though procedures are available to study velocity – dependent problems. However, Newtonian mechanics is applicable for both conservative and non – conservative forces.

3.21. IMPORTANCE / COMMENTS OF LAGRANGIN FORMULATION:

1. The Lagrangian formulation can be used to obtain the equations of motion of a system in any set of co – ordinates, not just the standard Cartesian co – ordinates, via the Euler Lagrange equation.

2. Lagrangian mechanics uses the energies in the system instead of forces. Energy is a scalar quality, hence we do not have to deal with vectors in this formulation, which makes it easier to deal with many complicated systems.
3. The Lagrange equations are used to derive the equations of motion of a solid mechanics, problems, including damping in matrix form.

4. It is only a alternative and equivalent formulation of Newton's formulation. The Newtonian and Lagrangian equations of motion are leads to second ordered differential equation. However the equations are obtained from different approach.

5. In Newtonian approach forces leads to acceleration the system this force cause the acceleration produced in the body. Therefore the accelerated motion is the effect of the force.

6. In Lagrangian approach to considered the Kinetic Energy (K.E) & Potential Energy (P.E) of the system. The concept of force does not entire into discussion directly. This is the difference between two formulations.

7. Another difference is that in some problem it may not be possible to know all the forces acting on the system, this is particularly true, if some constraints are introduced into the system but the expression for K.E & P.E may be given.

CHAPTER – 4

VARIATIONAL PRINCIPLES

In chapter 3, Lagrange's equations of motion were derived from D'Alembert's principle which is a differential principle. In this chapter the basic laws of mechanics are obtained from an integral principle known as Hamilton's Variational Principle. In this procedure, Lagrange's equations of motion are obtained from a statement about the value of the time integral of the Lagrangian between times t_1 *and* t_2. *In D'Alembert's principle, we considered the instantaneous state of the system and virtual displacements from the instantaneous state. However, in the following Variational procedure, an infinitesimal virtual variation of the entire motion from the actual one is considered.*

4.1. HAMILTON'S PRINCIPLE (Hamilton's Variational Principle)

According to Hamilton's Variational principle a particle moving from point A to B such that the line integral of Lagrangian with respect to time (say between t_1 & t_2) should be extremum (Either maximum or minimum)

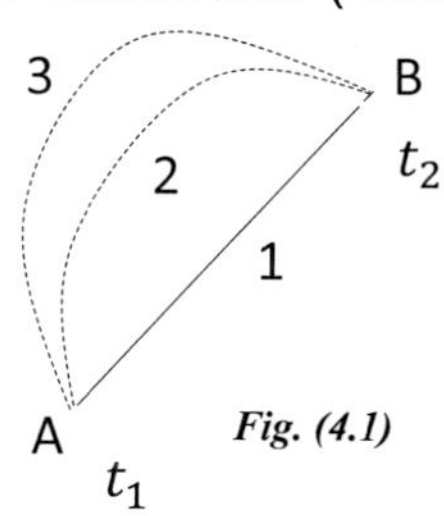

Fig. (4.1)

Mathematically it can be represented by

$$I = \int_{t_1}^{t_2} Ldt = J \; extremum \qquad ---------- (1)$$

Where

J is the extremum value of the time integral of the Lagrangian and known as Hamilton's principle function for the path

I is known as Action Integral

L is known as Lagrangian function L = T – V (K.E – P.E)

So, from equation (1) variation in time integration can be written as (i.e., δI must be zero)

$$\delta I = \delta \int_{t_1}^{t_2} Ldt = 0 \qquad ---------------- (2)$$

Where δ is the variation symbol

This principle helps us to distinguish the actual path from the neighbouring paths.

4.2. DEDUCTION OF HAMILTON'S PRINCIPLE

Let us consider that the conservative Holonomic dynamical system moves from P to Q, where P and Q are initial and final configurations of the system at times t_1 and t_2 respectively. Let PRQ be the actual path and $PR'Q$, $PR''Q$ the two neighbouring paths out of infinite number of possibilities.

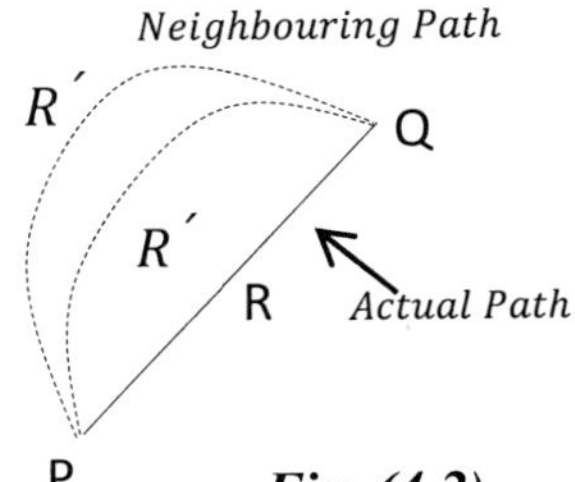

Fig. (4.2)

For the deduction of Hamilton's principle the following two conditions must be satisfied:

(i) δt must be equal to zero at end points, i.e., at t_1 the particle must be at P and at t_2 the particle must be at Q.

(ii) δr must be equal to zero at end points, i.e., the points P and Q are fixed in space.

Let the system be acted upon by a number of forces represented by F. Let the particle of the system acted upon by force F_i acquire acceleration $\ddot{r}_i$, so that we have

$$F_i = m_i \ddot{r}_i$$

From D'Alembert's principle, we have

$$\sum_i (F_i - m_i \ddot{r}_i)\, \delta r_i = 0$$

$$\sum_i F_i\, \delta r_i - \sum_i m_i \ddot{r}_i\, \delta r_i = 0 \quad \text{--------------- (1)}$$

But $$\ddot{r}_i \delta r_i = \frac{d}{dt}(\dot{r}_i . \delta r_i) - \dot{r}_i . \frac{d}{dt}(\delta r_i) \quad \text{--------------- (2)}$$

If there is a little variation along the actual and neighbouring paths, we have

$$\delta r_i = r_i' - r_i (Say)$$

Then $\frac{d}{dt}(\delta r_i) = \frac{d}{dt}(r_i' - r_i)$

$$\frac{d}{dt}(\delta r_i) = \frac{dr_i'}{dt} - \frac{dr_i}{dt}$$

$$\frac{d}{dt}(\delta r_i) = \delta\left(\frac{dr_i}{dt}\right)$$

$$\frac{d}{dt}(\delta r_i) = \delta(\dot{r}_i) \qquad \text{-------------- (3)}$$

Here primes have been used for neighbouring paths.

Put the value of equation (3) in equation (2) we get

$$\ddot{r}_i \delta r_i = \frac{d}{dt}(\dot{r}_i . \delta r_i) - \dot{r}_i . \delta(\dot{r}_i) \qquad \text{-------------- (4)}$$

Using above equation (1) becomes

$$\textstyle\sum_i F_i \, \delta r_i - \sum_i m_i \left[\frac{d}{dt}(\dot{r}_i . \delta r_i) - \dot{r}_i . \delta(\dot{r}_i)\right] = 0$$

$$\textstyle\sum_i F_i \, \delta r_i - \sum_i m_i \left[\frac{d}{dt}(\dot{r}_i . \delta r_i) - \frac{1}{2} \, \delta \dot{r}_i^{\,2}\right] = 0$$

$$\textstyle\sum_i F_i \, \delta r_i + \sum_i \frac{1}{2} m_i \delta \dot{r}_i^{\,2} = \sum_i \frac{d}{dt}(m_i \dot{r}_i . \delta r_i) \qquad \text{-------------- (5)}$$

But $\sum_i F_i \, \delta r_i$ = work done by the force F_i during displacement $\delta r_i = \delta W \; (say)$

And $\sum \frac{1}{2} m_i r_i^2$ = Kinetic energy of the system = T

Therefore, the equation (5) becomes

$$\delta W + \delta T = \textstyle\sum_i \frac{d}{dt}(m_i \dot{r}_i . \delta r_i)$$

Integrating above expression between the limits t_1 and t_2, we get

$$\int_{t_1}^{t_2} (\delta W + \delta T) dt = \int_{t_1}^{t_2} \textstyle\sum_i \frac{d}{dt}(m_i \dot{r}_i . \delta r_i) \, dt$$

$$= \textstyle\sum_i \int_{t_1}^{t_2} d\,(m_i \dot{r}_i . \delta r_i)$$

$$= \textstyle\sum_i [m_i \dot{r}_i . \delta r_i]_P^Q$$

$= 0 \; since \; \delta r_i = 0$ at the end points P & Q

For a conservative system, we know $\delta W = -\delta V$,

Where V is the Potential Energy

$\therefore \quad \int_{t_1}^{t_2}(-\delta V + \delta T)dt = 0$

or $\quad \int_{t_1}^{t_2} \delta(T - V)dt \quad$ i.e., $\int_{t_1}^{t_2}(T - V)dt = 0$

or $\quad \delta \int_{t_1}^{t_2} Ldt = 0$

or $\quad \int_{t_1}^{t_2} \boldsymbol{Ldt = J}$ **extremum**

This is the Hamilton's Principle.

4.3. DERIVATION OF LAGRANGE'S EQUATION FROM HAMILTON'S PRINCIPLE

(Deduction of Lagrange's Equation of Motion from Hamilton's Principle)

According to Hamilton's Variational principle

$$\delta I = \delta \int_{t_1}^{t_2} Ldt = 0$$

$$\delta I = \delta \int_{t_1}^{t_2}[T(q_k, \dot{q}_k) - V(q_k)]\, dt = 0$$

$$\delta I = \int_{t_1}^{t_2}[\delta T(q_k, \dot{q}_k) - \delta V(q_k)]\, dt = 0$$

$$\begin{bmatrix} We\ know\ that \\ f = f(x, y, z) \\ \delta f = \frac{\partial f}{\partial x}\delta x + \frac{\partial f}{\partial y}\delta y + \frac{\partial f}{\partial z}\delta z \end{bmatrix}$$

$$\delta I = \int_{t_1}^{t_2}\left[\left(\frac{\partial T}{\partial q_k}\delta q_k + \frac{\partial T}{\partial \dot{q}_k}\delta \dot{q}_k\right) - \frac{\partial V}{\partial q_k}\delta q_k\right] dt = 0$$

$$\delta I = \int_{t_1}^{t_2}\left[\left(\frac{\partial T}{\partial q_k}\delta q_k - \frac{\partial V}{\partial q_k}\delta q_k\right) + \frac{\partial T}{\partial \dot{q}_k}\delta \dot{q}_k\right] dt = 0$$

$$\delta I = \int_{t_1}^{t_2}\left(\frac{\partial T}{\partial q_k}\delta q_k - \frac{\partial V}{\partial q_k}\delta q_k\right) dt + \int_{t_1}^{t_2}\frac{\partial T}{\partial \dot{q}_k}\delta \dot{q}_k\, dt = 0$$

Taking summation on both sides to include all the particles of system

$$\int_{t_1}^{t_2}\sum_k\left(\frac{\partial T}{\partial q_k} - \frac{\partial V}{\partial q_k}\right)\delta q_k\, dt + \int_{t_1}^{t_2}\sum_k\frac{\partial T}{\partial \dot{q}_k}\delta \dot{q}_k\, dt = 0$$

$$\int_{t_1}^{t_2} \sum_k \left(\frac{\partial T}{\partial q_k} - \frac{\partial V}{\partial q_k}\right) \delta q_k \, dt \; + \int_{t_1}^{t_2} \sum_k \frac{\partial T}{\partial \dot{q}_k} \delta\left(\frac{dq_k}{dt}\right) dt \; = 0$$

$$\int_{t_1}^{t_2} \sum_k \left(\frac{\partial T}{\partial q_k} - \frac{\partial V}{\partial q_k}\right) \delta q_k \, dt \; + \int_{t_1}^{t_2} \sum_k \frac{\partial T}{\partial \dot{q}_k} \frac{d}{dt}(\delta q_k) \, dt \; = 0 \qquad \text{---------- (1)}$$

Integrating 2nd term of equation (1) by product rule

$$\left[\begin{array}{c} We\ know\ that \\ \int uv\, dt = u \int v\, dt - \int \frac{du}{dt} \int v\, dt\, dt \end{array}\right]$$

$$\int_{t_1}^{t_2} \sum_k \frac{\partial}{\partial q_k}(T - V)\delta q_k \, dt + \left[\sum_k \frac{\partial T}{\partial \dot{q}_k} \int_{t_1}^{t_2} \frac{d}{dt}(\delta q_k) dt - \int_{t_1}^{t_2} \frac{d}{dt} \sum_k \frac{\partial T}{\partial \dot{q}_k} \int \frac{d}{dt} \delta q_k \, dt\, dt\right] = 0$$

$$\int_{t_1}^{t_2} \sum_k \frac{\partial}{\partial q_k}(T - V)\delta q_k \, dt \; + \left[\sum_k \frac{\partial T}{\partial \dot{q}_k} \, (\delta q_k)_{t_1}^{t_2} - \int_{t_1}^{t_2} \frac{d}{dt} \sum_k \frac{\partial T}{\partial \dot{q}_k} \, (\delta q_k) dt\right] = 0$$

$$\int_{t_1}^{t_2} \sum_k \frac{\partial}{\partial q_k}(T - V)\delta q_k \, dt \; + \left[\sum_k \frac{\partial T}{\partial \dot{q}_k} \, \{\delta q_k(t_2) - \delta q_k(t_1)\} - \int_{t_1}^{t_2} \frac{d}{dt} \sum_k \frac{\partial T}{\partial \dot{q}_k} \, (\delta q_k) dt\right] = 0$$

$$\int_{t_1}^{t_2} \sum_k \frac{\partial}{\partial q_k}(T - V)\delta q_k \, dt \; + \left[\sum_k \frac{\partial T}{\partial \dot{q}_k} \, \{0 - 0\} - \int_{t_1}^{t_2} \frac{d}{dt} \sum_k \frac{\partial T}{\partial \dot{q}_k} \, (\delta q_k) dt\right] = 0$$

$$\int_{t_1}^{t_2} \sum_k \frac{\partial}{\partial q_k}(T - V)\delta q_k \, dt \; + \left[0 - \int_{t_1}^{t_2} \frac{d}{dt} \sum_k \frac{\partial T}{\partial \dot{q}_k} \, (\delta q_k) dt\right] = 0$$

$$\int_{t_1}^{t_2} \sum_k \frac{\partial}{\partial q_k}(T - V)\delta q_k \, dt \; - \int_{t_1}^{t_2} \frac{d}{dt} \sum_k \frac{\partial T}{\partial \dot{q}_k} \, (\delta q_k) dt \; = 0$$

$$\sum_k \int_{t_1}^{t_2} \left[\frac{\partial}{\partial q_k}(T - V) - \frac{d}{dt}\frac{\partial T}{\partial \dot{q}_k}\right] \delta q_k \, dt = 0 \qquad \text{------------ (a)}$$

$$\frac{\partial}{\partial q_k}(T - V) - \frac{d}{dt}\frac{\partial T}{\partial \dot{q}_k} = 0 \qquad \text{------------- (2)} \qquad \left[\begin{array}{c} We\ know\ that \\ If\ \int f(x)\, dx = 0 \\ Then\ f(x) = 0 \end{array}\right]$$

We know that $\quad \frac{\partial}{\partial \dot{q}_k}(T - V) = \frac{\partial T}{\partial \dot{q}_k} - \frac{\partial V}{\partial \dot{q}_k}$

$$\frac{\partial}{\partial \dot{q}_k}(T - V) = \frac{\partial T}{\partial \dot{q}_k} - 0$$

$$\frac{\partial}{\partial \dot{q}_k}(T - V) = \frac{\partial T}{\partial \dot{q}_k} \qquad \text{--------------- (3)}$$

Put the value of equation (3) in equation (2) we get

$$\frac{\partial}{\partial q_k}(T - V) - \frac{d}{dt}\frac{\partial}{\partial \dot{q}_k}(T - V) = 0$$

$$-\frac{\partial}{\partial q_k}(T-V)+\frac{d}{dt}\frac{\partial}{\partial \dot{q}_k}(T-V)=0$$

$$\frac{d}{dt}\frac{\partial}{\partial \dot{q}_k}(T-V)\ -\frac{\partial}{\partial q_k}(T-V)=0$$

$$\boldsymbol{\frac{d}{dt}\frac{\partial L}{\partial \dot{q}_k}-\frac{\partial L}{\partial q_k}=0}$$

Hence Proved

4.4. MODIFIED HAMILTON'S PRINCIPLE

According to Hamilton's principle, we have

$$\delta J=\delta\int_{t_1}^{t_2}L(q,\dot{q},t)dt=0 \qquad \text{---------------- (1)}$$

The relation between Lagrangian and Hamiltonian is given by

$$H(q,p,t)=\textstyle\sum_k P_k\dot{q}_k-L(q,\dot{q},t)$$

$$L(q,\dot{q},t)=\textstyle\sum_k P_k\dot{q}_k-H(q,p,t)$$

Then, equation (1) becomes

$$\delta J=\delta\int_{t_1}^{t_2}[\textstyle\sum_k P_k\dot{q}_k-H(q,p,t)]dt=0 \qquad \text{--------------- (2)}$$

This is called *Modified Hamilton's principle.*

4.5. DERIVATION OF HAMILTON'S EQUATIONS FROM MODIFIED HAMILTON'S PRINCIPLE (i.e., Variational Principle)

Consider two paths PRQ and $PR'Q$ out of infinite number of possibilities as shown in Fig. 4.3. In this case δ variation comprises independent variations of both q_k and p_k at constant time t. The difference between the two paths for the given value of t, may be described by introducing parameter α common to all points of the path of integration in phase space.

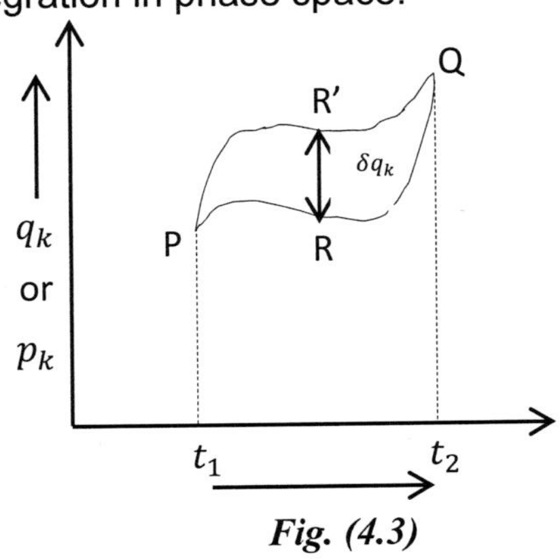

Fig. (4.3)

If $\bar{q}_k$ and $\bar{p}_k$ are the values of q_k and p_k for the varied path $PR'\ Q$ we have

$$\bar{q}_k = q_k + \delta q_k$$

$$\bar{q}_k = q_k + \frac{\partial q_k}{\partial \alpha}\delta\alpha$$

$$\bar{q}_k = q_k + \eta_k \delta\alpha \qquad \text{-------------- (1)}$$

and

$$\bar{p}_k = p_k + \delta p_k$$

$$\bar{p}_k = p_k + \frac{\partial p_k}{\partial \alpha}\delta\alpha$$

$$\bar{p}_k = p_k + €_k \delta\alpha \qquad \text{-------------- (2)}$$

Where $\frac{\partial q_k}{\partial \alpha} = \eta_k$ and $\frac{\partial p_k}{\partial \alpha} = €_k$ -------------- (3)

The variation in q_k and p_k are given by

$$\left.\begin{aligned} \delta q_k &= \eta_k \delta\alpha \\ \delta p_k &= €_k \delta\alpha \end{aligned}\right\} \qquad \text{--------------- (4)}$$

If $\delta\alpha = 0$, then $\delta q_k = \delta p_k = 0$, so that $\bar{q}_k = q_k$ and $\bar{p}_k = p_k$, i.e., that varied path $PR'\ Q$ coincides with the actual path PRQ.

As the times at end points are not varied so that

$$\left.\begin{aligned} \eta_k(t_1) &= \eta_k(t_2) = 0 \\ €_k(t_1) &= €_k(t_2) = 0 \end{aligned}\right\} \qquad \text{----------- (5)}$$

In terms of parameter α, Hamilton's Principle becomes

$$\delta J = \frac{\partial J}{\partial \alpha}\delta\alpha$$

$$\delta J = \delta\alpha \, \frac{\partial}{\partial \alpha}\int_{t_1}^{t_2}\left[\sum_k p_k \dot{q}_k - H(q,p,t)\right]dt \qquad \text{----------- (6)}$$

As the times at the end points are not varied and hence are not the functions of α, so that the differentiation and integration may be interchanged.

Then equation (6) may be written as

$$\delta\alpha \int_{t_1}^{t_2}\frac{\partial}{\partial \alpha}\left[\sum_k p_k \dot{q}_k - H(q,p,t)\right]dt$$

or $\delta\alpha \int_{t_1}^{t_2}\sum_k\left[\frac{\partial p_k}{\partial \alpha}\dot{q}_k + \frac{\partial \dot{q}_k}{\partial \alpha}p_k - \frac{\partial H}{\partial q_k}\frac{\partial q_k}{\partial \alpha} - \frac{\partial H}{\partial p_k}\frac{\partial p_k}{\partial \alpha}\right]dt = 0$ ------- (7)

But $\int_{t_1}^{t_2} \frac{\partial \dot{q}_k}{\partial \alpha} p_k dt = \int_{t_1}^{t_2} p_k \frac{d}{dt}\left(\frac{\partial q_k}{\partial \alpha}\right) dt$

$$\int_{t_1}^{t_2} \frac{\partial \dot{q}_k}{\partial \alpha} p_k dt = \left[p_k \frac{\partial q_k}{\partial \alpha}\right]_{t_1}^{t_2} - \int_{t_1}^{t_2} \dot{p}_k \frac{\partial q_k}{\partial \alpha} dt \quad \begin{bmatrix} Integrating \\ by\ parts \end{bmatrix}$$

$$\int_{t_1}^{t_2} \frac{\partial \dot{q}_k}{\partial \alpha} p_k dt = [p_k \eta_k]_{t_1}^{t_2} - \int_{t_1}^{t_2} \eta_k \dot{p}_k\, dt \qquad [From\ (3)]$$

$$\int_{t_1}^{t_2} \frac{\partial \dot{q}_k}{\partial \alpha} p_k dt = p_k \eta_k(t_2) - p_k \eta_k(t_1) - \int_{t_1}^{t_2} \eta_k \dot{p}_k\, dt$$

$$\int_{t_1}^{t_2} \frac{\partial \dot{q}_k}{\partial \alpha} p_k dt = 0 - \int_{t_1}^{t_2} \eta_k \dot{p}_k\, dt$$

$$\int_{t_1}^{t_2} \frac{\partial \dot{q}_k}{\partial \alpha} p_k dt = -\int_{t_1}^{t_2} \eta_k \dot{p}_k\, dt \qquad \text{-------------- (8)}$$

Now put the values of equation (3) and (5) in equation (7) we get

$$\delta\alpha \int_{t_1}^{t_2} \left[\text{€}_k \dot{q}_k - \eta_k \dot{p}_k - \frac{\partial H}{\partial q_k}\eta_k - \frac{\partial H}{\partial p_k}\text{€}_k\right] dt = 0$$

or $\int_{t_1}^{t_2} \sum_k \left[\text{€}_k \dot{q}_k \delta\alpha - \eta_k \dot{p}_k \delta\alpha - \frac{\partial H}{\partial q_k}\eta_k \delta\alpha - \frac{\partial H}{\partial p_k}\text{€}_k \delta\alpha\right] dt = 0$

or $\int_{t_1}^{t_2} \sum_k \left[\left(\dot{q}_k - \frac{\partial H}{\partial p_k}\right)\text{€}_k \delta\alpha - \left(\dot{p}_k + \frac{\partial H}{\partial q_k}\right)\eta_k \delta\alpha\right] dt = 0$

or $\int_{t_1}^{t_2} \sum_k \left[\left(\dot{q}_k - \frac{\partial H}{\partial p_k}\right)\delta p_k - \left(\dot{p}_k + \frac{\partial H}{\partial q_k}\right)\delta q_k\right] dt = 0$ ------ (9)

But the variation δq_k and δp_k are independent of each other; the integral (9) is satisfied only when the coefficient of δp_k and δq_k vanish separately, i.e., when

$$\dot{q}_k - \frac{\partial H}{\partial p_k} = 0 \text{ and } \dot{p}_k + \frac{\partial H}{\partial q_k} = 0$$

$$\frac{\partial H}{\partial p_k} = \dot{q}_k \text{ and } \frac{\partial H}{\partial q_k} = -\dot{p}_k \qquad \text{------------- (10)}$$

Which are the required Hamilton's equations.

4.6. PRINCIPLE OF LEAST ACTION

Action: The time integral of twice the K.E (T) is known as action.

Mathematically represented by

$$A = \int_{t_1}^{t_2} 2T dt$$

$$A = \int_{t_1}^{t_2} (L + H) dt$$

$$A = \int_{t_1}^{t_2} \sum_k p_k \dot{q}_k\, dt$$

We Know that

$$H = T + V \;\&\; L = T - V$$

$$H + L = (T + V) + (T - V)$$

$$H + L = 2T$$

$$As\ H = \sum_k p_k \dot{q}_k - L$$

$$H + L = \sum_k p_k \dot{q}_k$$

According to principle of least action for a conservative system

$$\Delta A = \Delta \int_{t_1}^{t_2} \sum_k p_k \dot{q}_k \, dt = 0$$

Where Δ represents a new type of variations of the path which allows time as well as position co – ordinates to vary.

Proof: (Deduction of Principle of Least Action)
The action integral *S* is given by

$$S = \int_{t_1}^{t_2} L \, dt \quad \text{-----------------} (1)$$

The Δ - variation of equation (1) is given by

$$\Delta S = \Delta \int_{t_1}^{t_2} L \, dt \quad \text{-----------------} (2)$$

We Know that

$$\Delta = \delta + \Delta t \frac{d}{dt} \quad \text{-----------------} (3)$$

Put the value of equation (3) in RHS of equation (2), we get

$$\Delta S = \left(\Delta = \delta + \Delta t \frac{d}{dt}\right) \int_{t_1}^{t_2} L \, dt$$

$$\Delta S = \delta \int_{t_1}^{t_2} L \, dt + \Delta t \frac{d}{dt} \int_{t_1}^{t_2} L \, dt$$

$$\Delta S = \int_{t_1}^{t_2} \delta L \, dt + \int_{t_1}^{t_2} \Delta t \frac{d}{dt} L \, dt$$

$$\Delta S = \int_{t_1}^{t_2} \delta L(q_k \, \dot{q}_k) \, dt + \int_{t_1}^{t_2} \Delta t \, dL$$

$$\Delta S = \int_{t_1}^{t_2} \left(\frac{\partial L}{\partial q_k} \delta q_k + \frac{\partial L}{\partial \dot{q}_k} \delta \dot{q}_k\right) dt + (L\Delta t)_{t_1}^{t_2} \quad \text{--------------} (4)$$

According to Lagrange's equation of motion we have

$$\frac{d}{dt}\left(\frac{\partial L}{\partial \dot{q}_k}\right) - \frac{\partial L}{\partial q_k} = 0$$

$$\frac{d}{dt}\left(\frac{\partial L}{\partial \dot{q}_k}\right) = \frac{\partial L}{\partial q_k}$$

$$\frac{d}{dt}(P_k) = \frac{\partial L}{\partial q_k} \quad \text{-------------------} (5)$$

Also $\delta \dot{q}_k = \frac{d}{dt} \delta q_k$ ------------------- (6)

Put the values of equation (5) & (6) in equation (4) we get

$$\Delta S = \int_{t_1}^{t_2} \left(\frac{d}{dt}(P_k)\delta q_k + P_k \frac{d}{dt}\delta q_k\right) dt + (L\Delta t)_{t_1}^{t_2}$$

$$\Delta S = \int_{t_1}^{t_2} \left(\frac{d}{dt} P_k \delta q_k\right) dt + (L\Delta t)_{t_1}^{t_2} \quad \text{-------------} (7)$$

$$\Delta q_k = \left(\delta + \Delta t \frac{d}{dt}\right) q_k \qquad \begin{pmatrix} We\ know\ that \\ \Delta = \delta + \Delta t \frac{d}{dt} \end{pmatrix}$$

$$\Delta q_k = \delta q_k + \Delta t \frac{d}{dt} q_k$$

$$\Delta q_k = \delta q_k + \Delta t\, \dot{q}_k$$

$$\delta q_k = \Delta q_k - \Delta t\, \dot{q}_k$$

Multiply with P_k on both sides we get

$$P_k \delta q_k = P_k \Delta q_k - P_k \Delta t\, \dot{q}_k \quad \text{----------------} (8)$$

Put the value of equation (8) in equation (7) we get

$$\Delta S = \int_{t_1}^{t_2} \sum_k \frac{d}{dt}(P_k \Delta q_k - P_k \Delta t\, \dot{q}_k)\, dt + (L\Delta t)_{t_1}^{t_2}$$

$$\Delta S = \sum_k [P_k \Delta q_k]_{t_1}^{t_2} - \sum_k [\Delta t\, P_k\, \dot{q}_k]_{t_1}^{t_2} + (L\Delta t)_{t_1}^{t_2}$$

$$\Delta S = 0 - \sum_k [\Delta t\, P_k\, \dot{q}_k]_{t_1}^{t_2} + (L\Delta t)_{t_1}^{t_2}$$

$$\Delta S = - \sum_k [\Delta t\, P_k\, \dot{q}_k]_{t_1}^{t_2} + (L\Delta t)_{t_1}^{t_2}$$

$$\Delta S = - \left[\sum_k (P_k\, \dot{q}_k - L)\Delta t\right]_{t_1}^{t_2} \qquad \begin{bmatrix} (\Delta q_k)_{t_1}^{t_2} = 0\ at \\ end\ points \end{bmatrix}$$

$$\Delta S = - [H\Delta t]_{t_1}^{t_2}$$

$$\Delta \int_{t_1}^{t_2} L dt = - [H\Delta t]_{t_1}^{t_2} \qquad \begin{pmatrix} We\ know\ that \\ S = \int_{t_1}^{t_2} L dt \end{pmatrix}$$

$$\Delta \int_{t_1}^{t_2} L dt = - \int_{t_1}^{t_2} H\, d(\Delta t)$$

$$\Delta \int_{t_1}^{t_2} L dt = - \Delta \int_{t_1}^{t_2} H\, dt$$

$$\Delta \int_{t_1}^{t_2} L dt + \Delta \int_{t_1}^{t_2} H\, dt = 0$$

$$\Delta \int_{t_1}^{t_2} (L + H) dt = 0$$

$$\Delta \int_{t_1}^{t_2} \sum_k p_k \dot{q}_k \, dt = 0$$

Hence Proved

4.7. LAGRANGE'S EQUATION FOR NON – HOLONOMIC SYSTEMS (Method of Lagrangian Undetermined Multipliers)

In the derivation of equation the Holonomic constraints do not appear until δq_k are considered to be independent of each other. In the non – Holonomic systems the generalized co-ordinates q_k are not independent of each other and they can not be reduced further by means of equation of constraints of the form

$$f(q_1, q_2 \dots, q_n, t) = 0$$

Let us consider non-holonomic system in which the constraints are represented by m equations of the form

$$\sum_{k=1}^{n} A_{jk} dq_k + B_{jk} dt = 0 \qquad \text{--------------- (1)}$$

Where k = 1, 2, 3, ..., m and n is the number of generalized coordinates, so that the number of degrees of freedom = $(n - m)$. According to Hamilton's principle, we have

$$\delta \int_{t_1}^{t_2} L dt = 0$$

Which from equation (a) of Section 4.3. gives

$$\int_{t_1}^{t_2} \sum_k \left[\frac{\partial L}{\partial q_k} - \frac{d}{dt}\left(\frac{\partial L}{\partial \dot{q}_k}\right) \right] \delta q_k \, dt = 0 \qquad \text{--------------- (2)}$$

It is to be noted that so far the constraints have not been accounted.

In Hamilton's principle the variation process involved is such that the time on each point of the path is held constant. Hence the restrictions on the variations due to constraints, from (1), are given by m equations of the form

$$\sum_k A_{jk} dq_k = 0 \qquad \text{--------------- (3)}$$

$$(Since\ dt = o)$$

These equations can be used to reduce the number of virtual displacements to independent ones. *The method for eliminating extra virtual displacements is called the method of Lagrangian undetermined multipliers.*

If equation (3) holds then we must have

$$\lambda_j \sum_k A_{jk} dq_k = 0 \qquad \text{-------------- (4)}$$

Where $\lambda_j (j = 1,2,3, \dots, m)$ are undetermined constants and are permitted to be functions of t, but not the other variables. Summing equation (4) and integrating the resulting equation from t_1 to t_2 we get

$$\int_{t_1}^{t_2} \sum_{i,j} \lambda_j A_{jk} dq_k \, dt = 0 \qquad \text{------------ (5)}$$

Comparing equation (2) and (5), we get

$$\int_{t_1}^{t_2} \sum_k \left[\frac{\partial L}{\partial q_k} - \frac{d}{dt}\left(\frac{\partial L}{\partial \dot{q}_k}\right) + \sum \lambda_j A_{jk}\right] \delta q_k \, dt = 0 \qquad \text{-------- (6)}$$

Here the displacements $\delta q_k{}'\, s$ are still not independent because they are connected by equation (3)

As there are $(n - m)$ degrees of freedom, the first $(n - m)$ of the co-ordinates $q_k{}'\, s$ may be considered to be independent while the last m are then fixed by equation (3).

So, for the values $\lambda_j{}'\, s$ are unspecified. Let us choose them such that

$$\frac{\partial L}{\partial q_k} - \frac{d}{dt}\left(\frac{\partial L}{\partial \dot{q}_k}\right) + \sum_j \lambda_j A_{jk} = 0 \qquad \text{------------ (7)}$$

For $k = n - m + 1, n - m + 2, \dots\dots, n$

With the values of λ_j given by (7), equation (6) may be written as

$$\int_{t_1}^{t_2} \sum_k \left[\frac{\partial L}{\partial q_k} - \frac{d}{dt}\left(\frac{\delta L}{\delta \dot{q}_k}\right) + \sum \lambda_j A_{jk}\right] \delta q_k \, dt = 0 \qquad \text{-------- (8)}$$

For $k = 1,2,3, \dots\dots, n - m$

In this equation $q_k{}'\, s$ are independent, so that this is only satisfied if the coefficients of δq_k separately vanish, i.e.,

$$\frac{\partial L}{\partial q_k} - \frac{d}{dt}\left(\frac{\partial L}{\partial \dot{q}_k}\right) + \sum_j \lambda_j A_{jk} = 0 \qquad \text{------------ (9)}$$

For $k = 1,2,3, \dots\dots, n - m$

From (7) and (9), we have

$$\frac{\partial L}{\partial q_k} - \frac{d}{dt}\left(\frac{\partial L}{\partial \dot{q}_k}\right) + \sum_j \lambda_j A_{jk} = 0 \qquad \text{For } k = 1,2,3, \dots\dots, n$$

or $\frac{d}{dt}\left(\frac{\partial L}{\partial \dot{q}_k}\right) - \frac{\partial L}{\partial \dot{q}_k} = \sum_j \lambda_j A_{jk}$ ------------- (10)

For $k = 1,2,3, \dots \dots, n$

Which represents complete set of Lagrange's equation for non – holonomic and conservative system.

4.8. HAMILTON'S PRINCIPLE FOR NON – HOLONOMIC SYSTEMS

Equations of constraints are in the form of algebraic expressions or in the differential form. If they are in the differential form, they can be directly incorporated into Lagrange's equations by means of Lagrange undetermined multipliers. We first discuss the Lagrange multiplier method for non – Holonomic systems and then for Holonomic systems.

With non-holonomic system, the generalized co-ordinates are not independent of each other. Consequently, for a virtual displacement consistent with constraints, the $\delta q_i{}' s$ are no longer independent. However, a straightforward treatment is possible if the equations of constraints are of the type

$$\sum_k A_{ik} dq_k + A_{it} dt = 0 \quad i = 1,2,3, \dots\dots, m \quad \text{-------- (1)}$$

Where the coefficients a_{ij} and a_{it} may be functions of the $q's$ and time. The quantity m indicates that there are m equations of this type. The constraint equation (1) for virtual displacement is given by

$$\sum_k A_{ik} \delta q_k = 0 \qquad i = 1,2,3, \dots\dots, m \quad \text{-------- (2)}$$

We can now use equation (2) to reduce the number of virtual displacements to independent ones by the Lagrange multiplier method. If equation (2) is valid, it is also true that

$$\lambda_i \sum_k A_{ik} \delta q_k = 0 \qquad i = 1,2,3, \dots\dots, m \quad \text{-------- (3)}$$

Where $\lambda_i (i = 1,2,3, \dots, m)$ are some undetermined quantities known as Lagrange's multipliers. In general, they are functions of the co-ordinates and of time t. In addition, Hamilton's principle is assumed to hold for non-holonomic systems. Proceeding as in equation (a) of Section 4.3.

$$\delta \int_{t_1}^{t_2} L\, dt = 0$$

$$\int_{t_1}^{t_2} \sum_k \left[\frac{\partial L}{\partial q_k} - \frac{d}{dt}\left(\frac{\partial L}{\partial \dot{q}_k}\right)\right] \delta q_k \, dt = 0 \quad \text{---------------- (4)}$$

Where the integration is with respect to time t_1 to time t_2. Summing equation (3) over k and then integrating with respect to time between time t_1 to time t_2. We have

$$\int_{t_1}^{t_2} \sum_k \sum_k \lambda_i A_{ik} \delta q_k \, dt = 0 \quad - \quad \text{------------ (5)}$$

Combining equations (4) and (5), we have

$$\int_{t_1}^{t_2} \sum_k \left[\frac{\partial L}{\partial q_k} - \frac{d}{dt}\left(\frac{\partial L}{\partial \dot{q}_k}\right) + \sum_k \lambda_i A_{ik}\right] \delta q_k \, dt = 0 \quad \text{------------- (6)}$$

In equation (6) the selection of $\lambda_i{'}s$ are at our disposal but the $\delta q_k{'}s$ are still not independent but they are connected by the m relations of equation (2). That is, out of the n co-ordinates $(n-m)$ may be selected independently and the remaining m are fixed by equation (2). The integrand in equation (6) can be split into two as

$$\int_{t_1}^{t_2} \sum_{k=1}^{n-m} \left[\frac{\partial L}{\partial q_k} - \frac{d}{dt}\left(\frac{\partial L}{\partial \dot{q}_k}\right) + \sum_k \lambda_i A_{ik}\right] \delta q_k + \int_{t_1}^{t_2} \sum_{k=n-m+1}^{n} \left[\frac{\partial L}{\partial q_k} - \frac{d}{dt}\left(\frac{\partial L}{\partial \dot{q}_k}\right) + \sum_k \lambda_i A_{ik}\right] \delta q_k \, dt = 0 \quad \text{-------------- (7)}$$

As the values of $\lambda_i{'}s$ are at our disposal, we now choose them to be such that the second term in equation (7) vanishes. That is,

$$\frac{\partial L}{\partial q_k} - \frac{d}{dt}\left(\frac{\partial L}{\partial \dot{q}_k}\right) + \sum_k \lambda_i A_{ik} = 0 \qquad k = (n-m+1), \ldots, n \quad \text{-------- (8)}$$

Now we are free to select the $n-m$ co-ordinates arbitrarily and therefore from equation (7)

$$\frac{\partial L}{\partial q_k} - \frac{d}{dt}\left(\frac{\partial L}{\partial \dot{q}_k}\right) + \sum_k \lambda_i A_{ik} = 0 \qquad k = 1,2, \ldots, (n-m) \quad \text{-------- (9)}$$

Combining equation (8) & (9), we get the complete set of Lagrange's equations for non – holonomic systems:

$$\frac{d}{dt}\left(\frac{\partial L}{\partial \dot{q}_k}\right) - \frac{\partial L}{\partial q_k} = \sum_{k=1}^{m} \lambda_i A_{ik} \qquad k = 1,2, \ldots, n \quad \text{------- (10)}$$

Now we have $n+m$ unknowns, the n co-ordinates $\delta q_k{'}s$ and the m Lagrange multipliers (λ_i). Equation (10) together with the equations of constraints, now being written as first order differential equations

$$\sum_k A_{ik}\dot{q}_k + A_{it} = 0 \qquad \text{-----------------} (11)$$

Constitute $(n + m)$ equations for $n + m$ unknowns.

Next we consider the physical significance of the undetermined multipliers. For that, let us remove the constraints on the system in such way that the motion is unchanged by the application of external forces Q_k. These external forces Q_k make the equations of motion of the system remain the same. Under the influence of these Q_k, the equations of motion are

$$\frac{d}{dt}\left(\frac{\partial L}{\partial \dot{q}_k}\right) - \frac{\partial L}{\partial q_k} = Q_k \qquad \text{------------} (12)$$

Which must be identical with equation (10). Hence, we can identify $\sum_k \lambda_i A_{ik}$ with Q_k, the generalized forces of constraint.

Next we consider a holonomic system in which there are more generalized co-ordinates than degrees of freedom. A holonomic equation of constraint

$$f(q_1, q_2, \dots, q_n, t) = 0 \qquad \text{------------} (13)$$

is equivalent to the differential equation

$$\sum_k \frac{\partial f}{\partial q_k} dq_k + \frac{\partial f}{\partial t} dt = 0 \qquad \text{------------} (14)$$

Which is identical in for to equation (1):

$$A_{ik} = \frac{\partial f}{\partial q_k} \quad \text{and} \quad A_{it} = \frac{\partial f}{\partial t} \qquad \text{------------} (15)$$

Thus, Lagrange's multiplier method can be used for holonomic constraints when

(i) It is inconvenient to reduce all the $q's$ to independent co-ordinates or

(ii) We desire to obtain the forces of constraint.

REFERENCES

1. Classical Mechanics - *G. Aruldhas*
2. Classical Mechanics - *Satya Prakash*
3. Classical Mechanics - *J.C. Upadhyaya*
4. Classical Mechanics - *H. Goldstein (Addison-wesley, 1950)*
5. Introduction to Classical Mechanics - *R.G. Takawale and P.S. Puranik (TMH, 1979)*